HYDRODYNAMICS

A STUDY IN LOGIC, FACT, AND SIMILITUDE

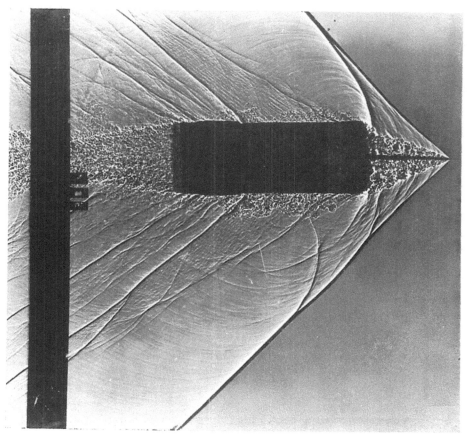

Common sense contradicted (*courtesy of Ballistics Research Laboratories, Aberdeen Proving Ground*).

HYDRODYNAMICS

A STUDY IN LOGIC, FACT AND SIMILITUDE

Revised Edition

by

GARRETT BIRKHOFF

PRINCETON, NEW JERSEY
PRINCETON UNIVERSITY PRESS
1960

TO RUTH

Preface

The present book is largely devoted to two special aspects of fluid mechanics: the complicated logical relation between theory and experiment, and applications of symmetry concepts. The latter constitute "group theory" in the mathematical sense.

The relation between theory and experiment is introduced in Chapters I and II by numerous "paradoxes," in which plausible reasoning has led to incorrect results. In Chapter III this relation is studied more closely in the special case of flows with "free boundaries."

Chapter IV is devoted to analysis of "modeling" and its rational justification. Theory and practice are compared (or contrasted!), and the origins of modeling in symmetry concepts are described as well, thus providing a transition between the two main aspects of fluid mechanics studied in this book.

The rest of the book centers around applications of group-theoretic ideas. These are shown in Chapter V to motivate a large proportion of the known exact solutions to problems involving "compressible" and "viscous" flows. In Chapter VI they are shown to yield the classical theory of virtual mass, as a special case of the modern geometrical theory of "homogeneous spaces."

The overall organization of material thus closely follows that of the first edition. It has, however, been very carefully revised in detail, and a number of interesting new developments of the past decade have been included.

The basic ideas were originally presented as Taft lectures at the University of Cincinnati in January 1947. These ideas matured and developed while the author was a John Simon Guggenheim Fellow in Europe in 1948. Similarly, the material in the first three chapters of the revised edition was presented as lectures at the Universidad Nacional de Mexico, and was carefully polished while the author was Visiting Professor there on sabbatical leave in 1958 under a Smith-Mundt grant. Finally, all the material was presented in a graduate course at Harvard University in 1959.

The author wishes to acknowledge partial support of preparation of the book by the Office of Naval Research.

The revision has been helped materially by suggestions from various colleagues and friends, and especially from Sydney Goldstein, Joseph Kampé de Fériet, Wallace Hayes, and Enzo Levi. Skillful typing by Laura Schlesinger and very careful proof-reading by Robert Lynch have also aided in its preparation. It is a pleasure to recall all this help.

Contents

HYDRODYNAMICS

A STUDY IN LOGIC, FACT, AND SIMILITUDE

I. *Paradoxes of Nonviscous Flow*

1. Rational hydrodynamics

Theoretical or "rational" hydrodynamics tries to predict real fluid motions (approximately), by solving *boundary-value problems* for appropriate systems of partial differential equations. In deriving these differential equations, one assumes Newton's Laws of Motion as axiomatic. One also assumes that the "fluid" (liquid or gas) considered is perfectly continuous, and that a well-defined pressure or other internal *stress* (force per-unit-area) is exerted across any element of surface, varying differentiably with position, time, and orientation, at least locally. Finally, one relates these stresses to the fluid motion by postulating various material constants (density, viscosity, etc.) and functional relations (adiabatic compression law, etc.). From such assumptions, mathematicians have derived systems of differential equations for various *idealized fluids* (incompressible nonviscous, compressible nonviscous, incompressible viscous, etc.).

To obtain well-determined or "well-set"[1] problems for such differential equations, one must also postulate appropriate boundary conditions, referring either to the initial state of motion, or to the movements of walls and obstacles bounding the fluid, or both. Rational hydrodynamics comprises the study of the boundary-value problems which result from coupling such boundary conditions with the differential equations for idealized fluids.[2]

It is easy for mathematicians to become persuaded that rational hydrodynamics is, in principle, infallible. Thus, Lagrange[3] wrote in 1788: "One owes to Euler the first general formulas for fluid motion . . . presented in the simple and luminous notation of partial differences. . . . By this discovery, all fluid mechanics was reduced to a single point of analysis, and if the equations involved were integrable, one could determine completely, in all cases, the motion of a fluid moved by any forces. . . ." Motivated by this faith, many of the greatest mathematicians have grappled with the problems of rational hydrodynamics, from Newton and Euler to the

---

[1] We use Hadamard's now classic terminology, whereby a boundary-value problem is called *well-set* when it has one and only one solution, depending continuously on the boundary values. See J. Hadamard, *Lectures on Cauchy's Problem*, Yale Univ. Press, 1923, p. 32.

[2] For simplicity we have ignored energy conservation and other thermodynamic considerations which may also be invoked (see § 14).

[3] *Mécanique analytique*, Paris, 1788, Sec. X, p. 271.

present day. Their analysis, often inspired by physical intuition, has suggested some of the most important concepts of the theory of partial differential equations: Green's function, vortex line, characteristic, zone of dependence, shock wave, eigenfunction, stability, and "well-set" problem, to name a few.

However, the boundary-value problems of rational hydrodynamics are exceedingly difficult, and progress would have been much slower if rigorous mathematics had not been supplemented by various *plausible intuitive hypotheses*. Of these, the following have been especially suggestive:

(A) Intuition suffices for determining which physical variables require consideration.

(B) Small causes produce small effects, and infinitesimal causes produce infinitesimal effects.

(C) Symmetric causes produce effects with the same symmetry.

(D) The flow topology can be guessed by intuition.

(E) The processes of analysis can be used freely: the functions of rational hydrodynamics can be freely integrated, differentiated, and expanded in series (Taylor, Fourier) or integrals (Laplace, Fourier).

(F) Mathematical problems suggested by intuitive physical ideas are "well set."

The preceding plausible assumptions are usually made tacitly, as a matter of course. The first two chapters of this book are largely devoted to an explicit examination of their fallibility.

2. Hydrodynamical paradoxes

Actually, Euler's equations have been integrated in many cases, and the results found to disagree grossly with observation, flagrantly contradicting the opinion of Lagrange. In hydrodynamics, such *apparent inconsistencies* between experimental facts and conclusions based on plausible arguments are called *paradoxes*, and the word will be used in this sense below.

These paradoxes have been the subject of many witticisms. Thus, it has recently been said [4] that in the nineteenth century, "fluid dynamicists were divided into hydraulic engineers who observed what could not be explained, and mathematicians who explained things that could not be observed." (It is my impression that many survivors of both species are still with us.)—Again, Sydney Goldstein has observed that one can read all of Lamb [7] without realizing that water is wet!

---

[4] Sir Cyril Hinshelwood, quoted by M. J. Lighthill in *Nature, 178* (1956), p. 343. See also [11], vol. 1, pp. 1–4.

It is now usually claimed that such paradoxes are due to the difference between "real" fluids having small but finite viscosity, and "ideal" fluids having zero viscosity.[5] Thus it is essentially implied that one can rectify Lagrange's claim,[3] by substituting "Navier-Stokes" for "Euler."

This claim will be discussed critically in Ch. II; it may well be correct in principle for *incompressible* viscous flow. However, taken literally, I think it is still very misleading, *unless* explicit attention is paid to the plausible hypotheses listed above, and to the lack of rigor implied by their use.

Though I do not know of any case when a deduction, both physically *and* mathematically rigorous, has led to a wrong conclusion, very few of the deductions of rational hydrodynamics can be established rigorously. The most interesting ones involve free use of one or more of Hypotheses (A)–(F).

This is illustrated by the Navier-Stokes equations themselves. They obviously fail to represent relativistic effects, molecular structure, and quantum effects, as well as such specific phenomena as ionization, electrostatic forces, impurities in suspension, condensation, etc., all of which can cause serious trouble, as will be shown below. Hence, from the beginning, Hypothesis (A) is freely invoked. In *compressible* flow even the meaning of bulk viscosity is open to question (§§22, 33).

I am not urging that Hypotheses (A)–(F) be no longer used in real hydrodynamics—even in pure mathematics, plausible arguments play a very important role.[6] In hydrodynamics, progress would hardly be possible without free use of such plausible hypotheses—and complete rigor is seldom possible. I am only insisting that plausible arguments must be checked, either by the discipline of rigorous proof (as in pure mathematics) *or* by experiment, before they can be accepted as scientifically established.

On the contrary, I think we should welcome the discovery of hydrodynamical paradoxes—recognizing frankly the inadequacy of existing mathematics (and logic) to analyze the complex wonders of Nature. Experience shows that man's imagination is far more limited than Nature's resources: as Pascal wrote, "l'imagination se lassera plûtot de concevoir que la nature de fournir."

Accordingly, the rest of this chapter will be devoted to an analysis of some paradoxes of classical hydrodynamics. In Ch. II attention will be turned to analogous (but less widely known) paradoxes of "modern" fluid dynamics.

[5] See [4], pp. 2, 48; [11], vol. 1, p. 3, and the introduction to vol. 2; Hunter Rouse, *Fluid Mechanics for Hydraulic Engineers*, McGraw-Hill, 1938, p. x.

[6] See G. Polya, *Induction and Analogy in Mathematics*, Princeton University Press, 1954. Paradoxes arise even in pure mathematics; see E. P. Northrop, *Riddles in Mathematics*, Van Nostrand, 1944.

3. Euler's equations

We begin by recalling the fundamental equations for *nonviscous* fluids, as defined by Euler and Lagrange. Let $u = u(x; t)$ denote the vector velocity of a fluid at position x and time t. Let $\rho(x; t)$ denote the fluid density, $g(x; t)$ the external gravity field, and $p(x; t)$ the fluid pressure.

If one admits the plausible Hypothesis (E) of §1 (ignoring the molecular structure of matter!), one easily shows that conservation of mass is equivalent to the partial differential equation

$$(1) \qquad \operatorname{div}(\rho u) + \frac{\partial \rho}{\partial t} = 0 \qquad \text{(Equation of Continuity)}.$$

If one denotes the "material derivative" with respect to time, for an observer moving with the fluid, by $D/Dt = \partial/\partial t + \sum u_k \, \partial/\partial x_k$, then one can rewrite (1) as

$$(1') \qquad \frac{D\rho}{Dt} + \rho \operatorname{div} u = 0.$$

The *incompressible* case corresponds to $D\rho/Dt = 0$, and hence to $\operatorname{div} u = 0$.

In a *static* fluid with $u = 0$ the fluid stress acts *normally* on any surface. This is the physical definition of a fluid; it is satisfied experimentally by many real substances.

Euler assumed that this law of hydrostatics applied also to moving fluids—i.e. to hydrodynamics. It is satisfied approximately in many fluid motions (except near the boundary). Thus ([4], p. 4), a change in velocity of 100 m.p.h. across a layer of air 0.01 in. thick, exerts a shear force of only one ounce per square foot, or less than 1/2000 of atmospheric pressure.

Continuous fluids which satisfy Euler's hypothesis are said to be *nonviscous*.[7] As shown by Cauchy, the stress in a nonviscous fluid must be the same in all directions (isotropic); the resulting scalar function $p(x; t)$ may be called the *pressure*. Further, conservation of momentum is equivalent to the vector partial differential equation

$$(2) \qquad \frac{Du}{Dt} + \frac{1}{\rho} \operatorname{grad} p = g \qquad \text{(Equation of Motion)}.$$

In order to get a self-contained system of partial differential equations in which all *time* derivatives can be expressed in terms of *space* derivatives,[8] one must supplement (1)–(2) by a further relation. In the rational

---

[7] Hypothesis (B) of §1 suggests that air and water can be considered as nonviscous fluids.

[8] Technically, so that initial conditions will define a *Cauchy* problem in the usual mathematical sense.

mechanics of homogeneous nonviscous fluids one usually assumes a relation connecting the density and the pressure,

$$(3) \qquad\qquad \rho = h(p) \qquad\qquad \text{(Equation of State)}.$$

Homentropic flows. Nonviscous fluids satisfying (3) may be called homentropic, and fluid motions satisfying (1)–(3), "homentropic flows." They occur in (approximately) homogeneous fluids, under conditions which are thermodynamically reversible. (By a "homogeneous fluid" we mean one having uniform composition, e.g. pure water or air.)

Thus, they occur typically in acoustics and in high-speed aerodynamics Rapid compression and expansion are typically adiabatic,[9] in the sense that heat conduction is negligible. Moreover the neglect of heat conduction is logically consistent with the neglect of viscosity in (2), because both heat conduction and viscosity are molecular effects. In an ideal gas with thermodynamic equation of state $p = RT$ and constant specific heat ratio $C_p/C_v = \gamma$, elementary considerations give, in adiabatic flow,

$$(3a) \qquad\qquad p = k\rho^\gamma,$$

the so-called *polytropic* equation of state for a calorically perfect gas. The limiting case $\gamma = 1$ corresponds to isothermal flow (infinite specific heat or, in an infinite isothermal reservoir, infinite conductivity).

Equation (3a) is accurate enough for many problems of gas dynamics; in air, $\gamma = 1.4$ nearly. But in liquids, one must approximate Equation (3) by $(p - p_v) = k\rho^\gamma$, where p_v is the cavitation (vapor) pressure; see §42.

A relation of the form (3) is also reasonable in fluids which are only slightly compressed (i.e. at speeds much less than that of sound, especially in liquids). In this case, one may write simply

$$(3b) \qquad\qquad \rho = \rho_0,$$

and speak of a *homogeneous, incompressible, nonviscous* fluid. However, in this case one can no longer express all time derivatives in terms of space derivatives.

4. Velocity potential

The fundamental equations (1)–(3) of Euler have various basic consequences with many important applications.

The most fundamental consequence is the Theorem of Helmholtz, valid for homentropic flow in conservative gravitational fields (i.e. if $g = \nabla G$). This theorem ([7], p. 36) asserts the invariance of the circulation $\Gamma =$

[9] We recall that Newton (*Principia Mathematica, Book II*, Sec. 8, Prop. 48) assumed Boyle's Law for isothermal flow, which led to an incorrect prediction of the velocity of sound. Newton's error was corrected by Laplace ([7], p. 477).

$\oint \sum u_k \, dx_k$ around any closed curve moving with the fluid—i.e. consisting of the same fluid particle at all times. Hence, if the fluid is initially at rest (e.g. flows from a static reservoir), and if the curve remains closed at all times, the circulation should always be zero. That is, there should exist a locally single-valued *velocity potential* $U(x; t)$, a scalar function of position such that

(4) $$u(x; t) = \nabla U = \text{grad } U.$$

Flows with this property are called (locally) *irrotational*. In a *simply connected* domain, such as the exterior of a solid in space or half the symmetric exterior of a circular cylinder in the plane, U must therefore be single-valued in the large.

For homentropic flows, in the absence of external gravity forces, irrotationality in the large (i.e. Eq. (4)) yields an integral of the Equations of Motion, the so-called Bernoulli Equation

(4′) $$\int \frac{dp}{\rho} = P(t) - \tfrac{1}{2}\nabla U \nabla U + \frac{\partial U}{\partial t}, \quad \rho = h(p).$$

In fact, the Equations of Motion (without gravity) are just the gradient of the relation (4′).

Incompressible flows. In the case of homogeneous incompressible fluids, the Bernoulli Equation (4′) can be generalized to include the effect of gravity. Namely, the gradient of

(5) $$p = P(t) - \rho_0 \left\{ \tfrac{1}{2}\nabla U \nabla U + \frac{\partial U}{\partial t} + G \right\}$$

is equivalent to the Equations of Motion *with* gravity, for irrotational incompressible flows. In this case, moreover, Eq. (1) simplifies to div $u = 0$, and hence to

(6) $$\nabla^2 U = 0.$$

Finally, on any impermeable solid boundary, clearly

(7) $$\frac{\partial U}{\partial n} = F(x; t)$$

is determined by the normal velocity of motion of this boundary.

Equations (6)–(7), for functions $U(x)$ single-valued in the large, define a classic problem of potential theory, the so-called *Neumann Problem*. As we shall see in §5 and Ch. VI, this problem has great importance for theoretical hydrodynamics. But first, let us note the implication of Hypothesis (F) of §1: the conjecture that the Neumann problem should have one and only one single-valued solution $U(x; t)$, for reasonably regular boundaries.

It is significant that this mathematical conjecture, suggested by hydrodynamical ideas, should have required more than 50 years for its rigorous demonstration. Today it is considered a fundamental theorem of pure potential theory ([6], pp. 310–11).

This theorem shows that, in an incompressible nonviscous fluid initially at rest, the velocity field at any instant depends only on the *instantaneous* velocity of the boundaries, and not on the previous history. Related theorems show that, as physical intuition would also suggest, the motion of any part of the boundary is instantly felt in all portions of the fluid: the signal velocity is infinite.

5. Steady irrotational flows

The case of steady (or permanent) flows, with $u = u(x)$, has an obvious special importance. The results of §4 made it plausible to 19th-century hydrodynamicists that, for a solid body moving with constant velocity for many diameters in an unlimited fluid of sufficiently low viscosity, initially at rest, one could write $U = U(x)$, $p = p(x)$, $g = g(x)$, etc.—all relative to axes fixed in the body, relative to which the fluid was moving with constant velocity a. The plausibility argument is clearly based on Hypothesis (B) of §1.

If this plausibility argument is accepted, then one can proceed mathematically as follows. In steady flows, with $U = U(x)$, the Equation of Motion (2) is equivalent after one space integration to[10]

$$(8) \qquad \frac{1}{2} \nabla U \nabla U + \int \frac{dp}{\rho} + G = \frac{p_0}{\rho_0}, \quad \text{or} \quad \sum u_i du_i + \frac{dp}{h(p)} + dG = 0.$$

This is called the Bernoulli equation for steady flow; in the incompressible case, it assumes the familiar simple form

$$(8^*) \qquad p = p_0 - \rho_0(\tfrac{1}{2} \nabla U \nabla U + G).$$

Similarly, the condition that the velocity of the body be $-a$, relative to the fluid at infinity, can be written

$$(9) \qquad \lim_{x \to \infty} \text{grad } U = a,$$

in incompressible *or* compressible flow. Finally, the requirement of steady flow implies that the flow boundaries are stationary. Hence it reduces the condition (7) of impermeability to

$$(7^*) \qquad \frac{\partial U}{\partial n} = 0 \quad \text{on the boundary.}$$

[10] If (3a) holds, then $\int dp/\rho = \gamma p/(\gamma - 1)\rho = c^2/(\gamma - 1)$.

In steady irrotational compressible flow, the Equation of Continuity (1) can still be expressed in terms of the single unknown function $U(x)$—*if* the effect of gravity is negligible, as it usually is at speeds high enough for compressibility to be appreciable.[11] (When gravity is not negligible, as in large-scale atmospheric motions, condition (9) cannot be satisfied, even though irrotational flow is self-consistent.)

Kinematics of homentropic flow. Under the conditions just described, setting $G=0$ in (8), we can write

$$\frac{dp}{d\rho} = c^2 = h' \frac{1}{h'(p)} = \frac{1}{J(\nabla U \cdot \nabla U)},$$

where J is the inverse of the function $2p - 2 \int h'(\rho)d\rho/\rho$. On the other hand, Eq. (1) with $\partial\rho/\partial t = 0$ implies ρ^{-1} div $(\rho u) = 0$, or

$$(10) \qquad \nabla^2 U = \frac{1}{c^2} \frac{\partial U}{\partial x_j} \frac{\partial U}{\partial x_k} \frac{\partial^2 U}{\partial x_j \partial x_k}.$$

Another form of (10) is

$$(10^*) \qquad \nabla^2 U = M^2 \sum \frac{u_j u_k}{q^2} \frac{\partial^2 U}{\partial x_j \partial x_k},$$

where the local "Mach number" $M = q/c$ is the ratio of the local flow speed q to the local sound speed c, and all coefficients $u_j u_k/q^2$ are one or less. Substituting for $1/c^2$ in (10) from the preceding display, we get ([10], p. 240)

$$(11) \qquad \nabla^2 U = J(\nabla U \cdot \nabla U) \sum \frac{\partial U}{\partial x_j} \frac{\partial U}{\partial x_k} \frac{\partial^2 U}{\partial x_j \partial x_k}.$$

The single partial differential equation (11), together with the boundary conditions (9) and (7*), reduce the steady *compressible* flow problem for a homentropic fluid of zero (small?) viscosity to another plausible boundary-value problem. Once this is solved, the pressure field can be deduced easily from (8).

Thus, we have reduced the steady flow problem to a purely kinematic problem. Given any mathematical solution of (11), (9), and (7*), if one defines the pressure field by (8) with $G=0$, then the Equation of Motion (2) will automatically be satisfied. Evidently the Neumann Problem of §4 is obtained as a limiting case, letting $c \to \infty$. Assumption (F) suggests much more: that the solution can be expanded in powers of M^2 (Rayleigh-Janzen method of [5], p. 275).

---

[11] In the incompressible case the effect of gravity is that of ordinary hydrostatic buoyancy, as shown in §21.

6. Reversibility Paradox

One of the most basic problems of fluid mechanics is the prediction of the force exerted on a solid in steady translation, with constant velocity a, through an otherwise stationary homogeneous fluid. If the solid moves parallel to a plane of symmetry, then this force can be resolved into a drag D, lift L, and moment M acting in this plane.

Lagrange could have seen, through a very simple consideration of *reversibility*, that the idealized boundary value problem of §5 could not possibly lead to a correct prediction of the resistance experienced by real solids moving through real fluids. The basic idea is the following (see [1]).

Definition 1. The *reverse* of a flow $u(x; t)$ is defined by $v(x; t) = -u(x; -t)$, both flows having the same pressure and density at corresponding points.

By direct substitution, one easily shows that the reverse of any flow satisfying Eqs. (1)–(3) also satisfies Eqs. (1)–(3)—though with reversed boundary conditions. In particular, we have the following:

Lemma. If $u(x)$ is a steady irrotational flow around a solid obstacle, with $u(\infty) = a$, then $v(x) = -u(x)$ is one, with $v(\infty) = -a$. Moreover $u(x)$ and $v(x)$ have the same pressure fields, and so the same D, L, and M.

This lemma is in *qualitative* contradiction with real hydrodynamics: in reality, a reversal of the direction of motion of a projectile ordinarily reverses D and L (though not M),[12] instead of leaving them unchanged.

It is instructive to analyze the preceding contradiction more closely. Until it has been shown that the boundary value problem of §5 is well-set,[1] one cannot conclude that its equations are erroneous. Perhaps some additional condition is needed. Indeed, as we shall see in §10, this may be true for *supersonic* flow (i.e. if the Mach number $M > 1$). To clarify this point, we make a further distinction:

Definition 2. A theory of rational hydrodynamics will be called: *incomplete* if its conditions do not uniquely determine the flow around a given obstacle; *overdetermined* if its conditions are mathematically incompatible; *false* if it is well-set but gives grossly incorrect predictions.

Theorem 1. Any reversible hydrodynamical theory is incomplete, overdetermined, or false as regards drag and lift predictions.

Subsonic case. In the *subsonic* case, $M < 1$ at least for sufficiently small Mach number; it has been shown recently[13] that the boundary-value

---

[12] Classical hydrodynamics does correctly predict the broadsiding tendency of axially symmetric obstacles; cf. [7], pp. 86, 170. (For bodies having fore-and-aft symmetry, flow reversal leaves L unchanged and reverses the sign of M.)

[13] D. Graffi, *J. Rat. Mech. Analysis*, 2 (1953), 99–106; D. Gilbarg, ibid., 233–51; D. Gilbarg and J. Serrin, ibid. 4 (1955), 169–75; L. Bers, *Comm. Pure Appl. Math.*, 7 (1954), 441–504; R. S. Finn and D. Gilbarg, ibid. 10 (1957), 23–64, and *Acta Math.*, 98 (1957), 265–76. This generalizes the result of §4 for $M = 0$ (the Neumann Problem).

problem defined by Eqs. (11), (9), and (7*) of §5 is well-set. Since it is of elliptic type, the mathematical solution $U(x)$ must even be *analytic*. Therefore, we conclude that the Euler-Lagrange equations define a *false* theory of steady subsonic flow.

Transonic case. The transonic case, in which $M < 1$ in some regions but the flow is locally supersonic, has been the subject of much controversy. Mathematical models of such transonic flows have been constructed, but they seem to resemble physical reality very slightly.[14] What is even more dramatic, *no* shock-free transonic flow is possible for some profiles. This Transonic Flow Paradox has recently been established by Cathleen Morawetz.[15] In the terminology of Theorem 1, this means that the transonic flow problem of rational (Euler-Lagrange) hydrodynamics may be *overdetermined*.

In §10 we shall see that the supersonic flow problem is typically incomplete—it is noteworthy that the different resolutions of the Reversibility Paradox, in the three preceding cases, conforms to the general mathematical theory of boundary-value problems of elliptic, mixed, and hyperbolic types.

7. D'Alembert Paradox

Better known and older than the Reversibility Paradox is the d'Alembert Paradox. According to this paradox, the assumptions of §5 imply $D = L = 0$. The cases of a unit circular cylinder (Fig. 1) and a sphere follow by symmetry from the explicit forms of the velocity potential:

$$(12a) \qquad U = a\left(x + \frac{x}{r^2}\right) \qquad \text{(cylinder)},$$

$$(12b) \qquad U = a\left(x + \frac{x}{2r^3}\right) \qquad \text{(sphere)},$$

and Bernoulli's Theorem (8*), with $g = 0$. In such cases of fourfold symmetry one can show $D = L = 0$, if the problem is well-set, by reversibility considerations alone ([1], p. 248).

More generally, the d'Alembert Paradox follows from the Reversibility Principle for any profile which has *central* symmetry—that is, which is mapped into itself under reflection in a fixed center of symmetry. The flow about a flat plate of Fig. 2a provides a case in point. Pressures on surface elements which correspond under this central symmetry are equal in magnitude and opposite in direction; hence their resultant is a pure couple.[12]

---

[14] R. von Mises, "Mathematical theory of compressible fluid flow," completed by Hilda Geiringer and G. S. S. Ludford, Academic Press, 1958, Art. 25.3.

[15] *Comm. Pure Appl. Math.*, *9* (1956), 45–68; *10* (1957), 107–31, and *11* (1958), 129–44.

In the general case, the demonstration is quite delicate and involves a sharp theorem about solutions of $\nabla^2 U = 0$ in the neighborhood of infinity. Namely, let $U(x)$ grow at infinity like the first power of $|x| = r$ at most. Then it can be shown that

$$U = a \cdot x + \phi(x),$$

where $\phi(x)$ is "regular at infinity" ([6], Ch. X, §8). By this we mean that $\phi(x)$ can be expanded in a convergent series, analogous to the series of negative powers of z in the Laurent expansion, whose terms are negative powers of r times spherical harmonics in the latitude and longitude. (For such solutions of $\nabla^2 U = 0$, the plausible Hypothesis (E) is thus supported by rigorous theorem.)

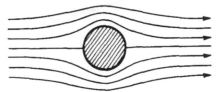

FIG. 1. Euler flow past a cylinder

8. Airfoil theory

Undiscouraged by the above paradoxes, scientists have succeeded in explaining drag and lift, at least qualitatively, within the framework of Euler's equations of motion. The trick is to avoid Hypothesis (D), as applied by Euler and Lagrange—this can be avoided using *discontinuous* and *many-valued* potentials. (Such functions are, to be sure, often considered pathological by "practical" men!)

As regards *drag*, one can postulate a stagnant *wake* ("dead water" region) with $U = 0$ behind an obstacle, extending to infinity as in Fig. 2c. This wake is separated from the main flow by a "free streamline" at constant pressure, across which the velocity $u = \nabla U$ changes discontinuously. This model will be analyzed in §39.

One can derive a theory of *lift* in two-dimensional flow by postulating a many-valued potential of the form

$$(13) \qquad\qquad U = \frac{\Gamma \theta}{2\pi} + \sum a_k x_k + \phi(x),$$

where
$$\nabla^2 \phi = 0, \quad \text{and} \quad \phi(x) = 0\left(\frac{1}{r^2}\right).$$

Here $\Gamma = \oint dU = \oint \sum u_k dx_k$, the integral being taken around the obstacle (airfoil), is the circulation defined in §4.

The introduction of the term $\Gamma\theta/2\pi$ in (13) sacrifices determinism—i.e.

it replaces the Neumann Problem of §4 by one which is not well-set. One can rescue an appearance of determinism if the airfoil has a sharp "trailing edge"—though not in general. In this case it is plausible to suppose that "the velocity is finite at the trailing edge" (Joukowsky Condition). This condition selects a unique Γ, and permits a prediction of drag and lift by the following Kutta-Joukowsky Theorem ([8], p. 188):

Theorem 2. In any plane flow of the form (13) we have $D=0$ and $L=\rho a\Gamma$, where $a=|a|$.

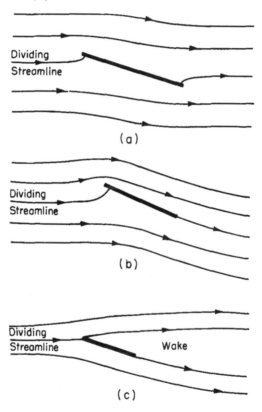

FIG. 2. Euler, Joukowsky, and Helmholtz flows past a flat plate

The special case $\Gamma=0$ yields the d'Alembert Paradox as a corollary.

One may define a steady, locally irrotational plane flow with circulation as a "Joukowsky flow" if it satisfies the Joukowsky Condition. The Joukowsky flow past a flat plate is sketched in Fig. 2b; its lift coefficient $C_L=2\pi \sin \alpha$, where α is the angle of inclination. The determination of the Joukowsky flow past a given profile with sharp "trailing edge" is a well-set boundary-value problem. Its solution in special cases (Joukowsky profile, Karman-Trefftz profile, etc.) constitutes a basic chapter in modern airfoil theory; the general theory was first given (with applications) by von Mises.[16] Its validity depends on the following theorem of pure mathematics, which enables one to transform the elementary Joukowsky flow (12a) around a unit circle into an incompressible Joukowsky flow around an arbitrary profile.

Fundamental Theorem of Conformal Mapping. There is one and only one complex analytic function

$$w = f(z) = kz + \sum_{0}^{\infty} c_k z^{-k}, \quad k > 0,$$

which maps the exterior of a unit circle one-one and conformally onto the exterior of a given simply connected domain.

This result has recently been extended to "quasi-conformal" maps[13]; it implies that there is one and only one subsonic Joukowsky flow at given Mach number $M < 1$ past an arbitrary profile with sharp trailing edge.

In the case of streamlined airfoils at small angles of attack, real flows are approximated quite well by ideal Joukowsky flows. Though the prediction of zero drag is obviously over-optimistic, the real lift is 75–95% of that predicted, and the lift/drag ratio can be as much as 50.

However, the Joukowsky Condition by no means gives a reliable theory of lift in general! Thus, in three-dimensional space the exterior of an airplane is clearly simply-connected. Hence, in space any locally irrotational flows must have a single-valued velocity potential U, with zero lift. If this were really true, flight would be impossible.

More subtle is the following Paradox of Cisotti.[17] Consider the Joukowsky flow around a flat plate, sketched in Fig. 2b. According to the Kutta-Joukowsky Theorem, the resultant force should be normal to the flow; since the stress is everywhere normal to the plate, it should be normal to the plate—a clear contradiction. As shown by Cisotti, the explanation is simple: there is a finite force on the trailing edge—due to the infinite negative pressure (suction) associated by (5) with the infinite velocity there. The paradox is thus associated with the failure of Hypothesis (E) of §1, and may be called a singular point paradox.

Unfortunately, experimental data do not confirm the variations in lift with the wing shape predicted by the Joukowsky theory. Thus, one has the following Fatness Paradox: theoretically C_L increases with the wing thickness; experimentally, it usually decreases.

[16] R. von Mises, *Zeits. Flugt. Motorluftschiffahrt*, 1917, pp. 157–63, and 1926, pp. 67–73 and 87–89. For an analysis of the facts, see R. von Mises, "Theory of flight," McGraw-Hill, 1945, Ch. VII.

[17] See G. Cisotti, *Rend. Accad. Lincei*, 5 (1927), 16–21, and 7 (1928), 17–19 and 538–43; also H. Pistolesi, ibid. *12* (1930), 409–11.

9. Magnus effect; drift

Golf and tennis players are familiar with the tendency of a spinning ball to deviate from its normal trajectory, in the direction towards which its forward side rotates. This tendency is called the Magnus effect. Following Rayleigh ([12], vol. 1, 343–6), the Magnus effect is usually explained qualitatively as follows.

Because of the spin, the local air-speed relative to ball is greater on the side where the rotation is backwards than where it is forward (see Fig. 3). Hence, by the Bernoulli equation (5), the pressure is less there—leading to a resultant force in the direction observed.

It is hard to make the above explanation quantitative because there is

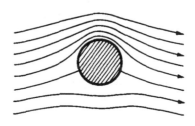

Fig. 3. Magnus effect

no obvious way to correlate the spin with circulation—even for a cylinder.[18] Prandtl made a valiant effort to predict at least the maximum lift L—which, he asserted, would occur when the circulation was just enough to allow a single stagnation point.[19] On this basis he predicted a maximum C_L of 4π. Recently this value has been transcended[20]—a fact which again shows the unreliability of non-rigorous arguments.

The inadequacy of existing explanations of the Magnus effect is shown even more strikingly by the following:

[18] As shown in §8, the case of a sphere leads to an even deeper paradox: by simple connectedness, $\Gamma = 0$.

[19] Prandtl's argument is summarized in ([4], §27); for a cylinder of radius c with relative translation a, $\Gamma = 4\pi ca$. For older experimental data, see [47], §239.

[20] W. M. Swanson, Final Report on Contract DA-33-019-ORD-1434, Case Inst. Technology, Dec. 31, 1956. With $V = \omega c = 17a$, $C_L = 14.7$ and "was still increasing at a constant rate." Recently, Prandtl's conclusion also has been disputed on theoretical grounds by M. B. Glauert, *Proc. Roy. Soc.* A242 (1957), 108–15.

Magnus Effect Paradox. At low rates of spin the actual deflection is in the opposite direction to that explained by Rayleigh (and observed by Magnus).[21]

To explain this Magnus Effect Paradox it seems that one must consider boundary-layer turbulence—a phenomenon which has so far defied mathematical treatment as a boundary-value problem. Thus, any correct explanation of the actual cross-force at low rates of spin must involve consideration of the Reynolds number.[22]

Analogous to the Magnus effect is the phenomenon of "drift." Artillerymen have known for over a century that spinning bullets tend to drift out of the vertical plane in which they are fired—and that this drift is in the

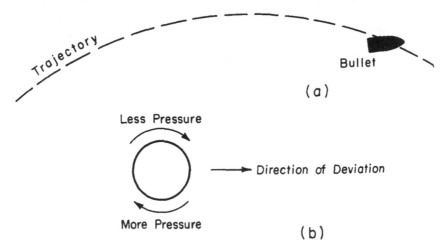

FIG. 4. Poisson's concept of Magnus effect

direction of rotation of the top of the bullet. However, this phenomenon was not correctly understood for many years.[23]

One false explanation was invented by the great mathematician Poisson. He reasoned that, due to its inertia, the bullet should lag behind the tangent to the trajectory, as sketched in Fig. 4a. There should thus be more pressure on the under side, and hence more friction there. As in Fig. 4b, this would lead to a drift in the observed direction. The fallacy of Poisson's argument is evident if one applies it to a spinning tennis ball: it predicts a deviation opposite to the usual Magnus effect!

The true explanation is as follows. One can verify by a quantitative study of gyroscopic stability that the stable position of the bullet is (with

[21] See [4], p. 504; the data (for spheres) are due to Maccoll. The data of [4], p. 546 suggest that the situation is similar for cylinders.

[22] See E. Krahn, *J. Aer. Sci., 23* (1956), 377–8.

[23] An interesting historical account is given in [3], Ch. X.

right-handed rifling) to the right of the trajectory tangent—not above it, as asserted by Poisson. The drift of the bullet is thus due directly to the aerodynamic cross-force and only indirectly to the spin.

This illustrates again the unreliability of qualitative arguments. A random explanation has 50 per cent chance of predicting the correct sign of a phenomenon!

10. Wave drag of thin wings

The d'Alembert Paradox cannot be extended to supersonic flow: even without considering viscosity, mathematical considerations lead one to expect a positive drag. In view of the Reversibility Paradox, this can only be because the boundary-value problem (for steady motion) defined by Euler's equations is not well set. We shall now indicate why this is so, beginning with linearized supersonic flow ("thin wing" theory).

Consider a family of time-independent compressible flows depending on a parameter δ, the wing thickness. We assume (Hypothesis (E) of §1) that the velocity potential can be written

$$(14) \qquad\qquad U = ax + \delta\phi(x, y, z) + 0(\delta^2).$$

Substituting into (10), and making the usual assumptions of perturbation theory, we get [24] as $\delta \downarrow 0$,

$$(14^*) \qquad\qquad (M^2 - 1)\phi_{xx} = \phi_{yy} + \phi_{zz}, \quad M = \frac{a}{c}.$$

The cases $M < 1$ of subsonic flow, $M = 1$ of sonic flow, and $M > 1$ of supersonic flow clearly correspond to partial differential equations of elliptic, parabolic, and hyperbolic type, respectively.[25] This simple observation already suggests that the boundary-value problem is only well set in the subsonic case.

In the case $\phi = \phi(x, y)$ of plane flow, the general solution of (14^*) is known, since d'Alembert, to be

$$(15) \qquad\quad \phi = F(x - \sqrt{M^2 - 1}\, y) + G(x + \sqrt{M^2 - 1}\, y),$$

where $F(r)$ and $G(s)$ are arbitrary functions.

To determine $F(r)$ and $G(s)$, we must use the condition (7), which reduces in stationary flow to $\partial U/\partial n = 0$ or

$$(15') \qquad\qquad\qquad \frac{\partial\phi}{\partial y} = a\eta'(x)$$

---

[24] See [2], §141; or [10], p. 245. For more detailed applications see [10], Ch. 8.

[25] This is equally true without linearization—but, in this case, M depends on position. Hence, flows may be transonic—and the corresponding differential equations of "mixed" type (elliptic in some regions and hyperbolic in others), as explained in §6.

for a "thin wing" bounded by the curve $y = \eta(x)$. We have replaced $\partial/\partial n$ by $\partial/\partial y$ in (15'), assuming that the slope $\eta'(x) \ll 1$. Actually, this hypothesis (or rather, $\eta'(x) \ll M$) is the basic assumption of thin-wing theory.

To escape the Reversibility Paradox, and to get a well-set problem, it is necessary to supplement (15)–(15') by an additional *irreversible* hypothesis, expressing the physical intuition that "waves travel downstream." If we place the thin wing along the x-axis, this hypothesis can be written

$$(15^*) \qquad \phi(x, y) = \begin{cases} F(x - \sqrt{M^2 - 1}\, y) & \text{if } y < 0 \\ G(x + \sqrt{M^2 - 1}\, y) & \text{if } y > 0. \end{cases}$$

Substituting into the Bernoulli equation (5), our equations imply the existence of a *wave pressure* given, in the perturbation approximation, by $p = \rho a^2 \eta'(x)$ on the upper wing surface $y = \eta(x)$, and by $p = \rho a^2 \tilde{\eta}'(x)$ on the lower surface $y = \tilde{\eta}(x)$. Taking the longitudinal stress component and integrating, we see that the predicted drag $D = \oint p\, dy$ is

$$(16) \qquad D = \rho a^2 \int (\eta' d\eta + \tilde{\eta}' d\tilde{\eta}) = \rho a^2 \int [\eta'^2 + \tilde{\eta}'^2]\, dx,$$

integrated over the length of the wing.

For sufficiently small slopes, the above formulas agree well with experiment ([10], pp. 346, 350) and evidently predict a positive supersonic "wave drag." Curiously, they agree with a very old quasi-empirical formula of Euler, which involved a universal constant factor supposed to be determined by experiment.[26]

11. Slender bodies of revolution

The application of the "Prandtl-Glauert" equation (14*) to supersonic flow past general "thin" or "slender" bodies is much too complicated to review here.[27] I shall only select a few examples to illustrate my general thesis: that predictions tend to be unreliable unless they are derived with mathematical and physical rigor.

A relatively simple problem concerns axial flow past solids of revolution (artillery projectiles at zero yaw). Von Karman and Moore[28] first drew the inference that the onset of wave drag should cause the resistance of a slender projectile to increase abruptly when $M = 1$—and estimated this increase on the basis of the simplifications of §10. Over a decade later Kopal extrapolated the argument to projectiles with yaw and showed that

---

[26] For applications to ballistic problems, see [3], §§12–16.

[27] See G. N. Ward, *Linearized Theory of Steady High-speed Flow*, Cambridge Univ. Press, 1955.

[28] *Trans. Am. Soc. Mech. Eng.*, *54* (1932), 303–10. For the modern "linearized" theory, see [10], §8.3; or G. N. Ward in [5], Ch. VIII, §15.

the simplified theory leads to several erroneous conclusions.[29] Especially, in the case of inclined cones, the cross-force calculated by the formulas of §10 decreases as M increases, whereas the true perturbation approximation makes it *increase* (Kopal Paradox).

It is now realized[28] that the simple linearized theory of §10 gives incorrect force predictions, even for slender bodies. The quadratic terms in the Bernoulli equation make a contribution to the pressure, for (supersonic) flow past slender bodies of revolution, which has the same order of magnitude as the linear term.[30]

In specific applications the simple linearized equations of §10 must be modified in many other ways.[31] Thus, for inclined wings of finite span, one must consider trailing vortex sheets. Moreover, in the case of round-headed solids of revolution such as spheres, the linearized boundary-value problem defined by (14*) and (15′) predicts fictitious singularities at the stagnation points (i.e. on the axis of symmetry). But the most conspicuous defect of "thin wing" theory is its failure to predict *shock waves*.

Shock waves are clearly visible as heavy lines in spark shadowgraphs of moving projectiles, such as that shown in the Frontispiece. In the case of cones and other sharp-nosed bodies at Mach numbers M sufficiently large, they are "attached" to the vertex, like the characteristics of solutions of linear hyperbolic differential equations. Otherwise, they are "detached," occurring in front of the projectile where the linearized theory would predict no disturbance whatever.

12. Earnshaw Paradox

The concept of a "shock wave" can also be deduced theoretically, starting from a simple paradox due to Earnshaw.[32] Our sense of hearing shows that sound travels nearly undistorted through long distances, with a constant velocity fixed by the air temperature. This observation makes it plausible that plane sound waves should travel without distortion or attenuation in an ideal nonviscous gas. This is however not true, as is shown by the

Earnshaw Paradox. Plane sound waves of finite amplitude and

[29] Z. Kopal, *Phys. Rev.*, *71* (1947), p. 474; [10], p. 377; detailed *Tables of Supersonic Flow Around Yawing Cones*, Mass. Inst. Technology (1947), esp. pp. xvi–xvii. Experimental data are given by M. Holt and J. Blackie, *J. Aer. Sci.*, *23* (1956), 931–6.

[30] This was apparently first shown by M. J. Lighthill, *Reps. Mem. Aer. Res. Comm.* *2,003* (1945); see also J. B. Broderick, *QJMAM*, *2* (1949), 98–120; and [10], p. 307.

[31] F. Ursell and G. N. Ward, *QJMAM*, *3* (1950), 326–48; S. Goldstein, *Proc. Int. Math. Congress*, Cambridge, 1950, vol. *2*, esp. pp. 288–9; M. C. Adams and W. R. Sears, *J. Aer. Sci.*, *20* (1953), 85–98.

[32] S. Earnshaw, *Phil. Trans.*, *150* (1860), 133–48. See also Stokes, *Phil. Mag.*, *33* (1848), p. 349; W. J. M. Rankine, *Phil. Trans.*, *160* (1870), p. 277; [12], vol. 5, p. 573 (or *Proc. Roy. Soc.*, *A84* (1910), 274–84); [7], §283; [2], §51.

stationary form are mathematically impossible in a gas which vibrates adiabatically.

Proof. We assume that the form of the sound waves is stationary, and that the train of waves moves with constant velocity normal to the wave fronts. Hence, if we change to axes moving with the waves, the fluid motion will be not only one-dimensional but *steady*. Choosing the x-axis as the direction of motion, we can thus write $\rho = \rho(x)$, $u = u(x)$, etc., and (neglecting gravity) Bernoulli's equation (8) will reduce to $u\,du + dp/\rho = 0$. Moreover, the Equation of Continuity (1) is simply $\rho u = \text{const} = C$, or $u = C/\rho$. Substituting back, we get

$$(17) \qquad \frac{-C^2 d\rho}{\rho^3} + \frac{dp}{\rho} = 0, \quad \text{or} \quad dp = \frac{C^2 d\rho}{\rho^2}.$$

Hence, such a wave motion is possible only if the fluid satisfies an Equation of State (3) of the special form

$$(18) \qquad p = p_0 - \frac{C^2}{\rho}.$$

There is no known gas whose adiabatic[33] equation of state has this special form.

13. Development of shocks

Like many other paradoxes, Earnshaw's contains the germ of an important truth. A more careful study of the equations involved establishes the principle that, in an adiabatic gas the denser portions of a wave of finite amplitude will gain on the rarer, ultimately overtaking them. The demonstration is as follows.

Let a denote the total mass of fluid to the left of a given point, so that $x(a, t)$ represents the position of the particle a at time t, and $\partial x/\partial a$ is the specific volume $\sigma = 1/\rho$. Then D/Dt in §3 becomes $(\partial/\partial t)_a$, $(\partial x/\partial t)_a = u$, and $(\partial^2 x/\partial t^2)_a$ is the material acceleration—the subscript a signifying that a is held constant. The Equation of Continuity (1') is automatically satisfied, since $D\rho/Dt = -\rho^2 \partial^2 x/\partial t \partial a$ and $\operatorname{div} u = \rho \partial^2 x/\partial a \partial t$. Moreover, the Equation of State (3) can be used to eliminate p through

$$(19) \qquad p = h\frac{1}{\sigma} = H(\sigma) = H\frac{\partial x}{\partial a}.$$

Consequently, neglecting gravity, (1)–(3) are equivalent to

$$(20) \qquad \frac{\partial^2 x}{\partial t^2} = -H'\frac{\partial x}{\partial a}\frac{\partial^2 x}{\partial a^2}.$$

---

[33] In fact, if (18) held, we would have $d^2p/d\rho^2 < 0$, which would contradict the Second Law of Thermodynamics.

Poisson discovered an important class of solutions of (20), given by the formula

$$(20^*) \qquad\qquad u = G(x-(c+u)t),$$

where $G(r)$ is an arbitrary function. These have been called *simple* waves ([2], p. 92); they are characterized by the property that $u = \int_{\sigma_0}^{\sigma} c(\sigma)\, d\sigma/\sigma$, where $c^2 = dp/d\rho = -\sigma^2\, dp/d\sigma$ is the squared speed of sound, which is a function of the specific volume σ. As Hadamard has shown, any plane wave entering a fluid (liquid or gas) initially at rest, from one side only, must be a simple wave.

Since $u = \int c(\sigma)\, d\sigma/\sigma$, (20^*) establishes a functional relation between u and σ—hence between u and c. Substituting back into (20^*), with $p = k\rho^\gamma$ $(\gamma > 1)$, one easily shows that the denser portions of gas gain on the rarer ([2], p. 96), and at a constant rate. Hence, in finite time, a *density discontinuity* or "shock wave" inevitably arises, violating Hypothesis (E) of §1 in a most striking way.

14. Thermodynamics of inviscid fluids

To interpret shock waves, one must keep in mind some basic thermo-dynamic ideas;[34] purely mechanical concepts do not suffice. Thus, one must consider the *internal energy* $E(p, T)$ of a fluid—even when it can be eliminated from the final equations, as in adiabatic flow. This quantity is involved in the conservation of energy, by the formula

$$(21) \qquad\qquad dQ = dE + p\, dV.$$

Here $p\, dV$ is the differential of work, in the absence of external forces.

One can define a *perfect gas* by Euler's equations and the thermo-dynamic equation of state $p = \rho RT$, where R is a material constant, and the formula $E = C_V T$ for internal energy, C_V being another constant (the specific heat at constant volume). In *isothermal* flow, $T = $ const., and $p = \rho RT$ imply (3a) with $\gamma = 1$. In *adiabatic* flow, heat is assumed to be transferred by convection only (no conduction or radiation); one then has $dQ = 0$ in (21). Per unit mass (so that $V = 1/\rho$), one then has $pV = RT$ and $E = C_V T = (C_V/R)pV$. Hence (21) gives

$$0 = dE + p\, dV = (C_V/R)V\, dp + (1 + C_V/R)p\, dV.$$

Setting $\gamma = (R + C_V)/C_V$, we get $dp/p = -\gamma dV/V = \gamma d\rho/\rho$. This implies the "polytropic" equation of state (3a), $p = k\rho^\gamma$.

---

[34] For an excellent and very relevant discussion of the thermodynamics of com-pressible fluids, see [10] or H. Liepmann and A. Roshko, *Elements of Gas Dynamics*, Wiley, 1957.

One can define a *perfect liquid* by Euler's equations and the condition $V = \text{const.}$ of incompressibility, Eq. (3b).

Rankine-Hugoniot Equations. By assuming conservation of mass, momentum, and energy, one can also predict the relation between the upstream values p_1, ρ_1, T_1 of pressure, density, and temperature, and the values p_2, ρ_2, T_2 just downstream of a normal shock. In a perfect gas, for example, these depend on a single parameter, the pressure ratio $P = p_2/p_1$ or *shock strength* $P - 1$. Thus ([10], p. 30):

$$\frac{\rho_2}{\rho_1} = \frac{P+\beta}{\beta P+1}, \qquad \beta = \frac{\gamma-1}{\gamma+1},$$

$$c = c_1\left[1 + \frac{(\gamma+1)(P-1)}{2\gamma}\right]^{1/2}.$$

The Reversibility Paradox, which would permit one to interchange 1 and 2 in the preceding formulas and to let $P < 1$, is avoided by an appeal to the Second Law of Thermodynamics. (In §13 the principle that dense portions of homentropic flows gain on the rarer portions also leads to this conclusion. It follows from the inequality $\gamma = (R+C_V)/C_V > 1$, which follows in turn from the physical positivity of R and C_V.)

The relations for oblique shocks can be easily deduced from those for normal shocks by use of moving axes (§67).

15. Breakers and bores

There is a remarkable analogy between long *irrotational* gravity waves in a shallow liquid of constant finite depth, and compression waves in an adiabatic gas with $\gamma = 2$. Long gravity waves of infinitesimal amplitude travel with constant speed $c = \sqrt{gh}$ without change of form, just as in the linearized supersonic flow approximation of §10. Long gravity waves of *finite* amplitude travel with a speed $\sqrt{gh}$ which increases with the local wave height. Consequently, the crest of any long wave in shallow water gains on the trough, in the way described in §13. The forward slope steepens continually, until it becomes vertical and the wave finally "breaks" under gravity.

This analogy was appealed to by Rayleigh,[35] to explain qualitatively the conversion of tidal waves travelling up estuaries into "bores." Such bores occur especially in gradually narrowing and shelving estuaries: the relative wave height of tidal waves is increased by the resulting concentration of total wave energy into a smaller cross-section and shorter wave length, equal to the product (12 hrs.) $\sqrt{gh}$.

[35] *Proc. Roy. Soc.*, *A90* (1914), 324–8; [7], Secs. 175–77, 182. See also D. Riabouchinsky, *Comptes Rendus*, *195* (1932), 998–9; [2], pp. 32–5. For Jeffreys' work, see V. Cornish, *Ocean Waves*, Cambridge, 1934, pp. 154–9.

Further mathematical insight was gained by the observation of Riabouchinsky,[35] who noted that the formula $c = \sqrt{g(h+y)}$, y being the local wave height, corresponded to the choice of $\gamma = 2$ in (3a). Soon after, Jeffreys[35] applied ideas similar to Rayleigh's, to the breaking of waves on sloping beaches. As waves advance into shallow water, the wave velocity decreases. This concentrates the wave energy into a shorter length, increasing the wave height and wave steepness even more. If the beach shelves gradually enough, the wave crest will again catch up to the trough, forming a "breaker" (surf).

Stoker[36] and others have tried to explain *quantitatively* the formation of "breakers and bores," using the above ideas. That is, they have tried to subsume these phenomena under Lagrangian rational hydrodynamics. However, it seems doubtful whether the fluid motion in real surf and tidal waves is sufficiently irrotational to make the model realistic. There is considerable vorticity in real surf and tidal waves, due to the backwash from preceding waves (undertow), currents, etc., and perhaps stratification due to sand in suspension. As a result, real breakers can "plunge," "spill," or "roll," while real bores can advance as isolated walls of water or in stages.[37] It seems unlikely that irrotational gravity waves will exhibit equal variety.

Moreover, it should be remembered that the ideal theory involves *two* parameters : the ratio h/λ of depth to wave-length, and the ratio h/R of depth to minimum radius of surface curvature R. For any *fixed* h and λ, waves of sufficiently small finite amplitude can travel without change of form, as shown in 1925 by Struik;[38] this apparent contradiction with the deductions of Rayleigh and Riabouchinsky may be called the Long Wave Paradox. The explanation is that Struik's construction refers typically to the case that h/R is comparable with h/λ, whereas the deductions of Rayleigh apply only to the case $h/\lambda \ll h/R \ll 1$.

16. Ferri Paradox

A much more recent paradox, due to Ferri,[39] concerns supersonic flow with "attached" shock wave past a tilted circular cone, whose axis is inclined at a "yaw" angle δ to the flow direction. As will be shown in §88,

---

[36] J. J. Stoker, *Water Waves, Interscience*, 1957, Secs. 10.7 and 10.10, and refs. given there.

[37] See M. Mason, on pp. 315–20 of *Gravity Waves*, Nat. Bu. Standards Circular 521, 1952; or Ch. III of Cornish, op. cit. in ftnt. 35.

[38] D. J. Struik, *Rendic. Lincei, 1* (1925), 522–7. For discussions of the Long Wave Paradox, see [7], loc. cit.; F. Ursell, *Proc. Camb. Phil. Soc., 49* (1953), 685–94; T. B. Benjamin and M. J. Lighthill, *Proc. Roy. Soc., A224* (1954), 448–60.

[39] A. Ferri, NACA Rep. 1045 (1951); see also M. Holt, *QJMAM, 7* (1954), 438–45.

Hypothesis (C) of §1 implies that such a flow will have *conical symmetry*. By this, we mean $u = u(\phi, \theta)$ in spherical coordinates.

Under central projection from the cone vertex, if similar streamlines are identified, the streamlines will therefore form a one-parameter family, which is sketched in Fig. 5. Except for the streamlines in the plane of symmetry, for which $\theta = 0, \pi$, all streamlines will tend to the limiting direction $(\alpha\text{-}\delta, \pi)$—that is, they will all feed into the straight streamline along the cone making the smallest angle $\alpha\text{-}\delta$ with the flow direction. But streamlines cutting the attached shock at different angles will have different entropies, by the Rankine-Hugoniot equations. Hence $u(\phi, \theta)$ will have a *singular point* at $(\alpha\text{-}\delta, \pi)$, again violating Hypothesis (E) of §1.

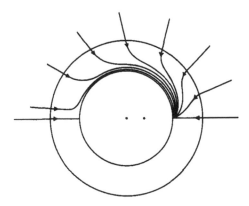

Fig. 5. Ferri paradox

Because of this singularity, expansions of $u(\phi, \theta)$ in powers of the yaw angle δ and Fourier series in θ are not justified. Hence Kopal's calculations of yaw effects (which were based on perturbation techniques assuming such expansions[40]) are not rigorous. Hence, neither is the Kopal Paradox (§11).

17. Triple Shock Paradox

The deduction of the Rankine-Hugoniot equations from conservation laws was mentioned in §14. These equations show that in a perfect gas the ratios p/p', ρ/ρ', T/T' of pressure, density, and temperature on the two sides of a stationary shock wave depend on a single parameter (the "shock strength" or Mach number—see [5], Ch. IV, §4). With the exceptions noted at the end of §14 these predictions are confirmed, and shocks of every strength can be observed.

Not so with multiple shocks. Corresponding mathematical predictions

[40] A. H. Stone, *J. Math. Phys.* MIT, 27 (1948), 67–81.

are possible in the case of double, "regularly" reflected shock waves (see Fig. 6a) and of "triple shocks" or "Mach Y" waves (see Fig. 6b), provided one assumes that in each relevant sector (1, 2, 3, 4) physical variables assume limiting values near the singular point at the shock "vertex." Again, many of these predictions are confirmed experimentally, so that the theory is very plausible.

However, in the case of "weak" shocks, regular reflections occur for angles of incidence somewhat greater than those allowed by theory, and *the predicted limits for triple shocks differ grossly from those observed.* This discrepancy, which may be called the Triple Shock Paradox, was ap-

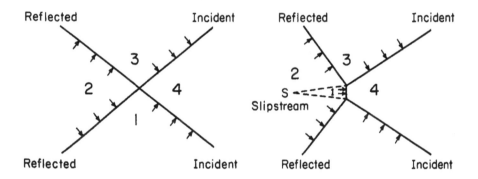

FIG. 6a. Regular reflection   FIG. 6b. Mach Y reflection

parently discovered by John von Neumann (1945). Repeated attempts have been made to resolve this paradox, which may be a "singular point paradox," due to an oversimplified guess as to the local behavior near the singular point. But no fully satisfactory explanation of it has yet been advanced.[41]

18. Value of Euler's equations

The preceding paradoxes show some limitations on the applicability of Euler's equations; but these equations are still the main tool of practical

[41] See W. Bleakney and A. H. Taub, *Revs. Mod. Phys.*, 21 (1949), 584–605; [2], p. 342; and [5], p. 144; H. Polachek and R. Seeger, *Phys. Rev.*, 84 (1951), 922–9; [3a], Sec. E, by the same authors, and refs. given there. For recent work, see R. G. Jahn, *J. Fluid Mech.*, 2 (1957), 33–48, and J. Sternberg, *Physics of Fluids*, 2 (1959), 179–206.

fluid mechanics. Thus, they permit one to calculate approximately: (i) the pressure distribution on the forward portions of obstacles, (ii) the lift of airplane wings, (iii) forces due to "cavity" motion (Ch. III) and jets, (iv) the hydrodynamic resistance to acceleration of a solid in fluid ("added mass"—see Ch. VI), (v) gravity wave propagation, including seiches and tides, (vi) sound propagation (acoustics), (vii) the pressure and flow distribution in supersonic nozzles, and (viii) supersonic drag.

In making such calculations, one should keep in mind the paradoxes described earlier, and also the great variety of flows compatible with nonviscous flow theory if vorticity is admitted. This variety is sometimes obscured in rational hydrodynamics by excessive emphasis on existence and uniqueness theorems—to prove which unrealistic assumptions are often made. It is emphasized in most books on "modern fluid dynamics" (e.g. in [4] and [17]), which point out from the beginning the possibility of separation—and such surprising experimental phenomena as boundary layers and turbulence.

The existence of such phenomena does not, however, necessarily relegate the purely mathematical theory of nonviscous fluids to an inferior position. Theory may suggest important possibilities which would be rejected by common sense as absurd.

For example, theory predicts the possibility of *wings of negligible drag*. Though this ideal has not yet been achieved, it has been the stimulus for much important work (see §29).

Again, the Reversibility Paradox suggests the possibility that a dead air region or "wake" might occur in *front* of a cylinder. Such a region would make it possible to establish the celebrated Taylor-Maccoll flow past a conical projectile (§85), in front of a flat-ended cylinder. This flow has the peculiar property that the cone wall is at constant pressure. According to the theory of "wakes" (Ch. III), the solid cone could be replaced by ideal nonviscous air at constant overpressure without destroying equilibrium. This suggests the mathematical possibility that in an ideal fluid supersonic flow past a flat disc might be possible, in which an invisible conical air barrier shields the disc from the onrushing air, greatly reducing the drag.

Common sense and intuition instantly reject this flow pattern as absurdly unstable. This rejection has the same logical basis as the rejection of an upstream "wake." It seems highly plausible that the effects of an obstacle should be on the downstream side.[42]

Yet, in this case common sense seems to be more fallible than mathematical deduction! Spark shadowgraphs taken at the Ballistic Research Laboratories in Aberdeen (see Frontispiece) indicate that a very thin

---

[42] The same intuition was used in §10 to select the preferred "simple" flow past a thin wing.

needle in front of the disc is actually capable of inducing something like this "absurd" flow.

In view of this and other examples, I think mathematicians need feel no inferiority about the soundness of deductive reasoning (as contrasted with "physical reasoning") in fluid mechanics, *provided* they are aware of the hydrodynamical paradoxes which experience has brought to light.

II. *Paradoxes of Viscous Flow*

19. Navier-Stokes equations

In spite of the great usefulness of the Euler-Lagrange equations, they are no longer generally accepted as providing an acceptable basis for rational hydrodynamics. Their place has been taken by the Navier-Stokes equations, whose derivation will now be briefly sketched.

These equations were first derived by Navier (1822) and Poisson (1829) from a simplified molecular model for gases, which led to a prediction of a positive *viscosity* $\mu > 0$, which was supposed to describe the molecular diffusion of momentum. However, the simple intermolecular force-laws assumed by these scientists are now recognized as hopelessly unrealistic, especially for liquids. Hence the continuum approach of St. Venant (1843) and Stokes (1845), which avoids these assumptions,[1] is today considered to be preferable philosophically. I shall begin by describing this continuum (or "macroscopic") approach.

This approach is based on the fundamental hypothesis that the pressure stresses considered by Euler must be supplemented by *viscous* stresses, which depend *linearly* on the *rate of strain*. A brief resume of the essential arguments involved follows.

The problem is to relate the (viscous) stress matrix $P = \|p_{ij}\|$ with the rate-of-strain matrix $\|\partial u_i / \partial x_j\|$. It was shown by Cauchy that in any material, if infinite rotational accelerations are to be avoided (Hypothesis (E) of §1), then the stress matrix must be symmetric: $p_{ij} = p_{ji}$. On the other hand, the rate-of-strain matrix is the sum of its skew component.

$$\frac{\|\partial u_i / \partial x_j - \partial u_j / \partial x_i\|}{2},$$

corresponding to rigid rotation, and its symmetric component

$$S = \frac{\|\partial u_i / \partial x_j + \partial u_j / \partial x_i\|}{2}.$$

Since rigid rotation does not involve any physical deformation, the symmetric matrix S expresses the true velocity of deformation.

Assuming isotropy, the principal axes of P and S must coincide (we

---

[1] For full refs, see [7], p. 577; for Stokes' ideas, see [13], vol. 1, p. 78 ff. and p. 182 ff. For a good modern discussion, see James Serrin, *J. Math. Mach.*, 8 (1959), 459–70.

recall the well-known theorem of algebra, according to which any symmetric matrix can be "diagonalized" by a rotation to suitable "principal" coordinate axes). Relative to these axes, it follows from our hypotheses of linearity and isotropy, after a little more consideration, that

$$(1) \qquad p_{ii} = p - \lambda \operatorname{div} \boldsymbol{u} - \frac{2\mu \partial u_i}{\partial x_i},$$

relative to principal axes, for suitable constants λ and μ. Moreover, by reflection symmetry,

$$p_{ij} = \frac{\partial u_i}{\partial x_j} + \frac{\partial u_j}{\partial x_i} = 0, \quad \text{if} \quad i \neq j.$$

Rotating axes into general position, we get

$$(1^*) \qquad p_{ij} = p - \lambda \operatorname{div} \boldsymbol{u} - \mu \left(\frac{\partial u_i}{\partial x_j} + \frac{\partial u_j}{\partial x_i} \right),$$

which are the basic equations to be applied below.

Unless incompressibility is assumed, it is very difficult to incorporate the Navier-Stokes equations into a well-set boundary-value problem whose conditions are physically consistent. To begin with, λ is often unknown (see §33). Stokes assumed tentatively that the "bulk viscosity" μ' vanished

$$(2) \qquad \mu' = \lambda + \tfrac{2}{3}\mu = 0,$$

which can be deduced from kinetic theory for monatomic gases.[2] One can also make a fallacious derivation of (2) by *defining* p as $(p_{11} + p_{22} + p_{33})/3$. The catch is, that we do not know that this "pressure" determines the density according to the thermodynamic equation of state $\rho = \rho(p, T)$ as determined from static measurements. If so, then (2) will follow ([7], p. 573); otherwise, we do not know how to relate the thermodynamic pressure to the stress tensor $\| p_{ij} \|$.

20. Real gases and liquids

Moreover, in real fluids λ and μ vary with the temperature T and pressure p; temperature variations are very important in lubrication, for example. At best, one can hope that $\lambda(p, T)$ and $\mu(p, T)$ are single-valued functions. To take this dependence into account mathematically, one needs to supplement the Navier-Stokes equations by the heat conduction equation, at least. This makes the boundary-value problem quite in-

---

[2] J. C. Maxwell, *Phil. Trans.*, *157* (1867), 49–88; S. Chapman and T. G. Cowling. *Mathematical Theory of Non-uniform Gases*, 2nd ed., Cambridge Univ. Press, 1952, Duhem showed that the Second Law of Thermodynamics requires $\mu' \geqq 0$.

tractable; even so, it is not physically exact, because radiation is neglected. (This neglect is very plausible by Hypothesis (B) of §1.)

Real fluid motions involve other physical effects not considered by Navier or Stokes. Thus, at hypersonic flow speeds relaxation effects, molecular dissociation, and ionization play an important role[3] in real gases. In dealing with satellites and re-entry problems, the would-be expert in fluid mechanics must supplement the Navier-Stokes equations by a familiarity with chemical kinetics. Similarly, the first supersonic wind-tunnels were plagued by condensation shocks due to water vapor in the air—another "hidden variable" ignored by the metaphysics of Navier and Stokes; see [3a, Sec. F.].

Sound attenuation in liquids and gases—which should certainly be predicted by any exact theory of compressible viscous fluids—is also found experimentally to be influenced strongly by molecular (relaxation) effects (see §33).

In view of all the complications mentioned above, it seems wise to concentrate attention on *incompressible* viscous fluids—even in this case, various anomalies are known. Thus, many liquids formed of long-chain molecules or containing clay suspensions are "non-Newtonian"—the current name used for liquids which do not satisfy the Navier-Stokes equations. The "superfluidity" of liquid He is another phenomenon which does not conform to the metaphysics of Stokes.[4]

21. Incompressible viscous fluids

Because of the complications described in §20, mathematicians have concentrated their attention on the Navier-Stokes equations for incompressible viscous fluids, assuming that μ and ρ can be taken as sensibly constant. Most experts believe that *rational hydrodynamics* based on these Navier-Stokes equations ordinarily provides a close approximation to *real hydrodynamics* when the Mach number M is very small, so that compressibility effects are negligible. To paraphrase Lagrange, they believe that "if the Navier-Stokes equations were integrable, one could determine completely all fluid motions at low Mach numbers" (cf. §1). In order to examine the basis for this belief, we will first put these equations into more convenient form.

Combining (1*) with incompressibility (Eq. (1') of Ch. I), we get

$$p_{ij} = p\delta_{ij} - \mu \left(\frac{\partial u_i}{\partial x_j} + \frac{\partial u_j}{\partial x_i} \right).$$

[3] See M. J. Lighthill in *Surveys in Mechanics*, Cambridge Univ. Press, 1956, 250–351; W. D. Hayes and R. F. Probstein, *Hypersonic Flow*, Academic Press, 1958.

[4] See F. London, *Superfluids*, Wiley, 1954, pp. 6–13; R. J. Donnelly, *Phys. Rev.*, 109 (1958), 1461–3, and refs. given there.

If one includes the terms in μ in the usual derivation of the equations of motion, one gets

$$(3) \qquad \frac{D\boldsymbol{u}}{Dt} + \frac{1}{\rho} \operatorname{grad} p = \boldsymbol{g} + \nu \nabla^2 \boldsymbol{u}, \quad \nu = \frac{\mu}{\rho},$$

instead of Euler's equation (3) of Ch. I. Combined with

$$(4) \qquad \operatorname{div} \boldsymbol{u} = 0,$$

Eq. (3) defines a (Newtonian) *incompressible viscous fluid*.

The effect of gravity on a solid submerged in such a fluid is easily accounted for, using the following principle.[5]

Theorem 1. In a viscous fluid with constant density ρ_0, the effect of gravity is equivalent to the superposition of a hydrostatic pressure $\rho_0 G$.

Proof. Setting $g_i = \partial G / \partial x_i$ in (3), where $-G$ is the gravitational potential, we get

$$(5) \qquad \frac{Du_i}{Dt} + \frac{1}{\rho_0} \operatorname{grad} \tilde{p} = \nu \nabla^2 u_i,$$

where $p = \tilde{p} + \rho_0 G$.

Caution. Note that the transformation of Theorem 1 does not conserve the usual "free surface" boundary condition $p = \text{const.}$ at a gas-liquid interface. Hence, it is useless when treating surface waves or cavities (Ch. III).

To define a well-set boundary-value problem from (3) and (4), we introduce the boundary condition of *no slip*[6] instead of Eq. (7) of Ch. I, namely:

$$(6) \qquad \boldsymbol{u}(\boldsymbol{x}) = \boldsymbol{0} \quad \text{on any fixed boundary.}$$

(At moving boundaries $\boldsymbol{u}(\boldsymbol{x}) = \boldsymbol{v}(\boldsymbol{x})$, the velocity of the moving boundary, whereas in the nonviscous case only the normal component of velocity is required to be continuous.)

We recall also the basic similarity principle

Theorem 2. Let $\boldsymbol{u} = \boldsymbol{f}(\boldsymbol{x}; t)$ satisfy (3), (4), and (6), with $\rho = \text{const.}$ If V, L, ν, V', L', ν', are constants such that $VL/\nu = V'L'/\nu'$, then

$$(7) \qquad \boldsymbol{u}' = \boldsymbol{g}(\boldsymbol{x}; t) = \left(\frac{V'}{V}\right) \boldsymbol{f}\left(\frac{L'\boldsymbol{x}}{L}, \frac{LV't}{L'V}\right)$$

also satisfies (3), (4), and (6), with ρ replaced by ρ' and p by p', where

$$(8) \qquad p' - \rho' G = \frac{\rho' V'^2}{\rho V^2} (p - \rho G).$$

---

[5] Th. von Karman, *Jour. Aer. Sci.*, 8 (1941), 337–56. In the nonviscous case the result goes back to d'Alembert, *Théorie de la résistance des fluides*, Arts. 48, 56, 91, and Avanzini (1807).

[6] The significance of this condition will be discussed in §34.

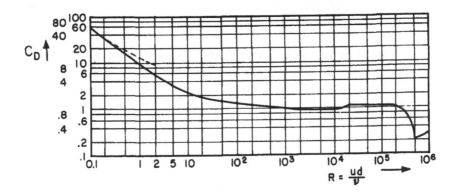

FIG. 7a. $C_D(R)$
(for cylinder)

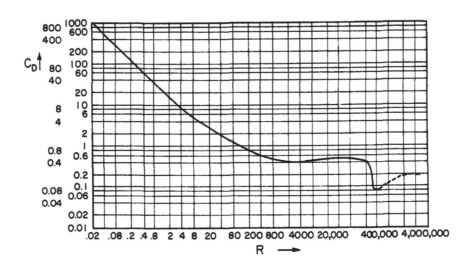

FIG. 7b. $C_D(R)$
(for sphere)

In other words, every change of scale (in space and time) which conserves the *Reynolds number* $VL/\nu = \rho VL/\mu = R$ carries incompressible flows which satisfy the Navier-Stokes equations into solutions of the same equations.

Corollary. If two steady flows satisfy the boundary-value problem (3)–(4)–(6) at the same Reynolds number, *and* if this boundary-value problem is mathematically well set, then they must have the same drag coefficient $C_D(R)$.

Now, classic experimental verifications of the *conclusion* of this corollary (though not necessarily of its hypotheses!) are depicted in Figs. 7a and 7b, which depict the drag coefficients of a cylinder (broadside) and a sphere, respectively. The existence of these remarkable curves could hardly have been suspected before the properties of viscosity had been given a clear mathematical formulation!

The ideas underlying Theorem 2 will be analyzed in detail in §71.

22. Non-analyticity Paradox

Lagrange constructed the first proof of the fact that in a nonviscous fluid the irrotationality of a molecule of fluid is permanent. Unfortunately, as shown by Stokes ([13], vol. 1, pp. 106–12), Lagrange's proof is fallacious. It applies equally well to regions of viscous fluids, in which such irrotationality is *not* permanent! The fallacy consists in assuming that the velocity and vorticity are *analytic* functions of time.

If this is assumed (by Hypothesis (E) of §1), then one can argue as follows. The fundamental equation (3) is equivalent (taking the curl of both sides) to

$$(9) \qquad \frac{D\boldsymbol{\xi}}{Dt} = \nu \nabla^2 \boldsymbol{\xi} + (\boldsymbol{\xi} \cdot \nabla) \boldsymbol{u}$$

in the vorticity $\boldsymbol{\xi} = \nabla \times \boldsymbol{u}$. Using "Lagrangian" independent variables $\boldsymbol{a}, t$, where $\boldsymbol{a}$ refers to a moving particle so that $\partial/\partial t$ ($\boldsymbol{a}$ fixed) is D/Dt, we can transform partial derivatives by

$$\frac{\partial}{\partial x_k} = \sum A_{kj}(\boldsymbol{a}, t) \frac{\partial}{\partial a_j}.$$

Substituting back into (9), we have

$$(10) \qquad \frac{D\xi_i}{Dt} + \sum \xi_k(\boldsymbol{a}, t) A_{kj}(\boldsymbol{a}, t) \frac{\partial u_i}{\partial a_j} = \nu \nabla^2 \xi_i + (\boldsymbol{\xi} \cdot \nabla) u_i.$$

Differentiating (10) successively with respect to t, for fixed $\boldsymbol{a}$, we get a sequence of explicit formulas for $D^n \xi_i / Dt^n$. One shows easily that every term in each formula contains as factor either ξ_i, $\nabla^2 \xi_i$, or one of

$$D\xi_i/Dt, \cdots, D^{n-1}\xi_i/Dt^{n-1}, \text{ etc.}$$

Therefore, by induction on n, assuming all functions to be infinitely differentiable, *all* $D^n\xi_i/Dt^n = 0$.

The effect of viscosity is through terms in spatial derivatives of the vorticity. In a nonviscous fluid, initial irrotationality $\xi(a, 0) = 0$ at any point $x(a, 0)$ guarantees that all $D^n\xi/Dt^n(a, 0) = 0$ at that point, whereas in the viscous case, irrotationality in an entire neighborhood of $x(a, 0)$ is required, to make all spatial derivatives of the vorticity vanish.

In either case, if $\xi(a, t)$ is *analytic* in t, then it vanishes identically—since all terms in its Taylor series (in t) do. This proves the following[7]

Non-analyticity Paradox. In order to acquire vorticity, a region of fluid initially at rest (or in irrotational motion) must have a vorticity which is a non-analytic function of time.

23. Existence and uniqueness

Before becoming certain about the adequacy of the Navier-Stokes equations to describe real (incompressible) fluid mechanics, we should know that they permit one to formulate physically natural boundary-value problems in ways which are "well-set" mathematically (cf. Thm. 2, Cor.). That is, we should have existence and uniqueness theorems, which have so far been proved only under restrictive hypotheses.

As regards the initial value problem, existence and uniqueness has been proved for plane and axially symmetric flows, under the restriction to finite total energy. The proof involves (9) which simplifies in plane flows to

$$(11) \qquad \frac{D\zeta}{Dt} = \nu\nabla^2\zeta, \qquad \zeta = \frac{\partial v}{\partial x} - \frac{\partial u}{\partial y}.$$

In space, however, even for finite total energy, existence only has been proved, and this only for limited time intervals.[8] Though the assumption of finite total energy can probably be relaxed—bounded velocity might suffice—E. Hopf[9] has shown that the initial value problem is not well set for the Navier-Stokes equations if the velocity is allowed to increase linearly and the pressure quadratically with distance from the origin.

For the steady flow problem, mathematical existence theorems have been proved[8] for both plane and space flows past general obstacles, but *not* uniqueness theorems. If the steady flows in question are unique, they are physically *unstable* at large Reynolds numbers; this is clear from the Turbulence Paradox (§25).

[7] P. Duhem, *Traite d'Énergétique*, vol. 2, p. 121; C. Truesdell, *Kinematics of Vorticity*, Indiana Univ. Press, 1954, §104.

[8] J. Leray, *J. de Math.*, *12* (1933), 1–82; ibid. *13* (1934), 331–418; *Acta Math.*, *63* (1934), 193–248; E. Hopf, *Math. Nachr.*, *4* (1951), 213–31. See also D. E. Dolidze, *Prikl. Mat. Meh.*, *12* (1948), 165–80, and *19* (1955), p. 764.

[9] E. Hopf, *J. Rat. Mech. Anal.*, *1* (1952), p. 107.

24. Poiseuille flow

As was true of Euler's equations in the time of Lagrange, the Navier-Stokes equations have been integrated in only a few cases. Agreement with experiment in these few cases is therefore of crucial importance.

One such case is furnished by flow through a long straight tube whose cross-section is a circle of constant radius c. We let x denote distance along the pipe, and r, distance from the axis of the pipe. In these cylindrical coordinates we let u_x, u_r and u_θ be the axial, radial, and angular components of velocity.

Theorem 3. The only solutions of (3), (4), and (6) which possess the symmetries of the hypotheses (of steady viscous flow in a circular tube) are the *Poiseuille* flows defined by

$$(12) \qquad u_x = a(c^2 - r^2), \qquad u_r = u_\theta = 0.$$

Proof. The hypothesis of "steady flow" means that $u = u(x, r, \theta)$, independent of t. The hypotheses of the problem are, moreover, invariant under reflection in every plane through the tube axis; a flow has these reflection symmetries if and only if $u_\theta = 0$, $u_x = f(x, r)$, $u_r = g(x, r)$. The hypotheses are also invariant under arbitrary translations along the pipe axis; since ν is assumed independent of pressure, so are (3) and (4). This translation symmetry is equivalent to $u_x = f(r)$, $u_r = g(r)$. Using (6), this implies div $u = d[rg(r)]/dr = 0$, whence $g(r) = C/r = 0$, since $g(r) = 0$ on the axis.

We now use (3), setting $g = 0$ by Theorem 1. Since $u_2 = u_3 = 0$, it gives $p = p(x)$. Considering the case $i = 1$ (x-component) we get

$$p'(x) = \mu \nabla^2 u_x = \mu[f''(r) + r^{-1}f(r)].$$

Since the left side is independent of r, the right side must be, too. Thus $(rf')' = rf'' + f' = kr$ for some constant k, and $rf' = \frac{1}{2}kr^2 + K$. This gives finite $u_x = f(r)$ at $r = 0$ only if $K = 0$; hence $f' = kr/2$ and $f(r) = \frac{1}{4}kr^2 + b$. To satisfy the "no slip" condition (6), we must have $u_x = a(c^2 - r^2)$, completing the proof of (12).

Substituting back into Eq. (3), we obtain the classic conclusion that if $Q = \pi a c^4/2$ is the volumetric discharge through the tube, the pressure gradient is

$$(13) \qquad -\frac{dp}{dx} = 4\mu a = \frac{8\mu Q}{\pi c^4}.$$

25. Turbulence Paradox

The experimental facts are most remarkable. Though the law (13) (Law of Poiseuille-Hagen) is observed for flow in capillary tubes, it breaks down completely for ordinary hydraulic pipes. More precisely, we can assert the following general

Turbulence Paradox. For flow in straight tubes, the Symmetry Hypothesis (C) of §1 is fulfilled if the Reynolds number $R < 1,700$, but not usually if $R > 10^4$. When $R > 10^4$, the observed flow usually has neither the temporal nor the spatial symmetry of the hypotheses and becomes turbulent.

The qualification "usually," in the preceding statements, refers to the possibility of avoiding turbulence by carefully fairing the inlet, polishing the walls, and having a laminar entering stream. With extreme care, turbulence has been avoided up to $R = 40,000$ in this way. But unless special care is taken, flow in pipes with $R > 2,000$ will be turbulent.

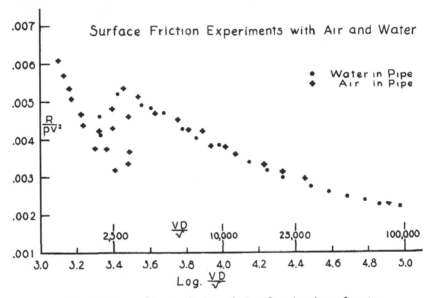

FIG. 8. Reynolds similarity of pipe flow in air and water

This is illustrated by the classical experimental data of Stanton and Pannell,[10] which are reproduced in Fig. 8. In a peculiar way they confirm the Navier-Stokes equations, by showing that the *critical Reynolds number* R_{crit} for transition to turbulence is about 1,700, the same for air as for water. This would follow rationally from Theorem 2 above. The opinion of most contemporary experts is that Poiseuille flow is simply *unstable* for $R > R_{crit}$, but that turbulent flow conforms to the Navier-Stokes equations. Though the similarity principle (7) of Theorem 2 does not imply the validity of the Navier-Stokes equations, their validity in turbulent flow is confirmed by experimental measurements of the rate of decay of homogeneous turbulence.[10a]

---

[10] *Phil. Trans.*, *A214* (1914), 119–24; see [4], Ch. VIII for more detailed studies of turbulent flow in pipes.

[10a] R. W. Stewart, *Proc. Camb. Phil. Soc.*, 47 (1951), 146–57.

Moreover, Hypothesis (C) of §1 is still fulfilled *statistically*. Using bars to denote averages, symmetry is exhibited by such formulas as

$$\overline{u_x} = F(r), \qquad \overline{u_r} = \overline{u_\theta} = 0, \qquad \overline{u_x^2} = G(r), \qquad \overline{u_r^2} = H(r),$$

and so on. Thus, a consideration of the experimental data suggests the concepts of *statistically* defined solutions of partial differential equations as a new and fascinating area for mathematical research. The study of such "stochastic differential equations" is now opening up new frontiers in pure analysis.

In spite of valiant attempts,[11] the observed instability of Poiseuille flow has defied mathematical analysis. It has even been conjectured[12] that Poiseuille flow is *stable* with respect to infinitesimal perturbations in perfectly smooth circular pipes. However, the instability of plane Poiseuille flow between parallel plates, even for two-dimensional perturbations if $R > 5{,}300$, is now fairly reliably established.[11] Hence this conjecture seems unlikely.

26. Other symmetry paradoxes

It is most peculiar to have Nature's love of symmetry limited to the range $R < 1{,}700$. Curiously, her supposed love of Least Action seems to have similar limits, since Poiseuille Flow involves much less expenditure of energy than turbulent flow.[13] To give some insight into her emotions, we consider other examples from hydrodynamics, in which a naïve application of Hypothesis (C) of §1 also yields incorrect results.

One interesting example is furnished by flow in pipes of noncircular cross-section. With small R, parallel flow is again observed; the velocity profile can be calculated ([7], §332); and Least Action reigns. At large R, the flow is again turbulent and not even statistically parallel: there are significant "secondary currents" into corners.[14]

Another case has been studied by Sir Geoffrey Taylor in a classic paper.[15] Consider a viscous fluid between two long coaxial cylinders, rotating with constant angular velocities ω and ω' in opposite directions. The problem described has (approximate) symmetry under translation along and rotation about the axis of the cylinders and is time-independent. There is just one solution of (3)–(4)–(6) which possesses these sym-

---

[11] See J. L. Synge, *Hydrodynamical stability*, Semicentennial publ. Am. Math. Soc., 1938, vol. *2*, pp. 227–69; [20], esp. §3.2.

[12] See [11], vol. 2, pp. 32–3; R. Comolet, *Comptes Rendus, 226* (1948), p. 2049; (also, *La Houille Blanche*, numéro special B (1949), p. 673). For theoretical arguments. see C. L. Pekeris, *Proc. Nat. Acad. Sci. USA, 34* (1948), 285–95.

[13] For technical theorems bearing on minimum energy expenditure, see [7], §344.

[14] [4], Sec. 161, where Nikuradse's data are reproduced.

[15] *Phil. Trans., A223* (1922), 289–93; see also [20], Ch. 2.

metries; it is called "Couette flow." At low ω, ω' this Couette flow is observed. At high Reynolds numbers Couette flow is replaced by an unsymmetrical flow which is, however, *not* turbulent. Roughly speaking, the symmetry in time is preserved, though spatial symmetry is not.

Again, consider a small air bubble, rising in still water under its own buoyancy. Under surface tension it will be nearly spherical—and in any case, there is no cause which is not symmetric about a vertical axis through the centroid of the bubble. Hence, by symmetry the bubble should rise vertically. Yet it is a striking fact[16] that if $R > 50$, such a bubble wobbles upwards in a vertical spiral instead!

An analogous phenomenon occurs in the wake behind a circular cylinder pulled broadside through a stream. In the range $50 < R < 500$ the wake consists of alternating vortices of opposite sign (Benard-Karman vortex street)—a phenomenon which will be analyzed in §56.

At first sight, such examples might seem to contradict Leibniz's metaphysical Principle of Sufficient Reason,[17] our Hypothesis (C). A deeper view is that, *although symmetric causes must produce symmetric effects, nearly symmetric causes need not produce nearly symmetric effects*: a symmetric problem need have no *stable* symmetric solution. This possibility is the real source of "symmetry paradoxes" (apparent violations of Hypothesis (C) of §1).

Now, wherever it is necessary to analyze the *stability* of a deduction, since an analysis of the stability of a mathematical deduction is far more complicated than the deduction itself, it is almost certain that the deduction will be made long before its instability can be tested. From this principle we may anticipate a continuing stream of symmetry paradoxes.

27. Boundary-layer theory

A fundamental question of fluid mechanics concerns the relation of solutions of Euler's equations for nonviscous fluid motion to those of the Navier-Stokes equations for fluids of vanishingly small viscosity. Mathematically, the problem concerns the asymptotic behavior of solutions of (3)–(4), as $\mu \downarrow 0$ (i.e. as $R \uparrow +\infty$). Since ships and airplanes normally involve Reynolds numbers in the range 10^6–10^9, the drag problem is also of enormous practical importance in this range.

It is easily seen that ordinary perturbation theory is not applicable to

---

[16] See H. Nisi and A. W. Porter, *Phil. Mag.*, *46* (1923), p. 754; [15], p. 294; P. J. Saltman, *J. Fluid Mech.*, *1* (1956), 249–75.

[17] A modern formulation of this has been given by George D. Birkhoff, Rice Institute Pamphlet 28 (1941), No. 1, pp. 24–50; *Collected Papers*, vol. 3, pp. 778–804. Curiously, formal systems of mathematical logic seem to have ignored this principle, whose applicability to physics was observed in 1894 by Pierre Curie (*Oeuvres Scientifiques*, Paris, 1908, esp. pp. 119, 215).

this problem, because the viscous term $\nu\nabla^2 u$ in (3) is of highest order, and so the perturbation of ν about the value $\nu=0$ is a *singular perturbation*.[18] In partial differential equations the type is ordinarily determined by the terms of highest order. Thus to suppress the terms of highest order will obliterate the type. Even for ordinary differential equations such as $\epsilon y''+y=0$, with the boundary conditions $y(0)=a$, $y(1)=b$, one gets completely different limiting behavior if one lets $\epsilon\uparrow 0$ from what one gets by letting $\epsilon\downarrow 0$.

Most contemporary discussions of the above singular perturbation problem start with Prandtl's idea that the vorticity is confined to a thin *boundary layer* of rapidly shearing fluid along each solid boundary, and a trailing wake of vorticity (often nearly a "vortex sheet") behind the body. Outside this boundary layer and wake, the flow is nearly irrotational, and Euler's equations are applicable.

For the boundary layer itself, Prandtl[19] constructed a model involving a self-consistent neglect of various terms. For two-dimensional flow he obtained (neglecting gravity)

$$(14) \qquad \frac{\partial u}{\partial t}+u\frac{\partial u}{\partial x}+v\frac{\partial u}{\partial y}+\frac{1}{\rho}\,p'(x) = \nu\frac{\partial^2 u}{\partial y^2}, \quad p = p(x),$$

plus the usual incompressibility condition (4), $\partial u/\partial x+\partial v/\partial y=0$. This is of *parabolic* type, and can be integrated step-by-step so long as $u>0$, relative to the boundary conditions $u(x,0)=v(x,0)=0$ on the fixed wall, and $u(x,\infty)=u_\infty(x)$ outside the boundary layer, where $u_\infty(x)$ is supposed given in terms of the pressure by the Bernoulli equation, $\rho u_\infty^2/2+p(x)=$ const. At the first point where $u(x,y)<0$ for positive y, the flow is supposed to *separate*.[20]

Various attempts have been made to rigorize Prandtl's somewhat intuitive derivation of (14); the most noteworthy is perhaps that of von Mises.[21] However, the theoretical facts are still far from clear. Besides an unresolved singularity at the leading edge, they involve the following

Boundary-Layer Paradox. Theoretical derivations of (14) assume that the ratio δ/x of boundary layer thickness δ to length x tends to zero. Experimentally, if $\delta/x<0.01$, the boundary layer becomes *turbulent*, and (14) is not satisfied.[22]

[18] This has been pointed out by Oseen ([9], p. 211) and others.
[19] *Proc. Third Int. Math. Congress Heidelberg* (1904), pp. 484–91, reprinted in [21].
[20] See [4], Ch. IV; or [23], pp. 94–9 and 222–4 for more complete discussions.
[21] *ZaMM*, *7* (1927), 425–7. For a heuristic discussion of the singularity in the flow near the leading edge of a flat plate, see G. F. Carrier and C. C. Lin, *Quar. Appl. Math.*, *6* (1948), 63–8.
[22] Cf. [23], pp. 32–3. Other difficulties in naively applying the theory are mentioned on pp. 110–12. For still other difficulties, see K. Stewartson, *J. Math. Phys.* MIT, *36* (1957), 173–91.

For example, the boundary layer remains laminar for only a few centimeters along ship hulls and on airplane wings during flight! This boundary-layer turbulence can be related to pipe turbulence, as we shall now see.

28. Paradoxes of Eiffel and Dubuat

The idea that the resistance D of a bullet should be a smoothly increasing function of its velocity v is very old. Thus, in many texts one finds "proofs" (by dimensional analysis, see §61) that D must be proportional to v at low speeds, and to v^2 at high speeds. It was therefore most surprising when Constanzi and Eiffel[23] discovered in 1912 the following

Eiffel Paradox. Near a critical Reynolds number $R_c \simeq 150,000$ the fluid resistance experienced by a sphere actually decreases as the velocity increases.

Two years later Prandtl showed that the drop in resistance depended on transition to *turbulence* in the "boundary layer" around the sphere, and could be induced by roughening the sphere *or* by making the incoming stream turbulent. In fact, one can correlate the $R_c \simeq 150,000$ of Eiffel's Paradox with the $R_{\text{crit}} \simeq 1,700$ for pipe turbulence, by treating the laminar boundary-layer thickness δ as analogous to the pipe diameter d. If $R = u_\infty x / \nu$ is the main-flow Reynolds number at a distance x from the leading edge, then $\delta(x) \simeq 4\sqrt{\nu x / u_\infty}$ by (14), and so the boundary-layer Reynolds number $R_\delta = u_\infty \delta / \nu \simeq 4\sqrt{u_\infty x / \nu}$. Hence $\delta / x \simeq 4/\sqrt{R}$, and the $R_c \simeq 150,000$ of Eiffel's Paradox corresponds roughly to $R_\delta \simeq 4\sqrt{R_c} \simeq 1,600$ —in good agreement with R_{crit} for turbulence in pipes. This discovery explains also the following older paradox.

Paradox of Dubuat. The resistance of a stick held in a stream flowing with velocity v is usually less than that of the same stick towed with velocity v through still water.

This paradox is especially interesting because the paradox seems at first sight to contradict a basic principle of Newtonian mechanics: the preservation of all laws under transformation to moving axes in uniform translation. Presumably, because he recognized this principle, Leonardo da Vinci[24] asserted the similarity of the two cases above—an assertion directly confuted by observation.

Today the reason seems obvious. Flowing streams are always more or less turbulent; this induces a drop in resistance for the same (mathematically unexplained) reason as it does for flow around a sphere, as shown by

---

[23] G. Eiffel, *Comptes Rendus*, 155 (1921), 1597-9. For Prandtl's explanation, see [11], vol. 2, Sec. 63—or *Gott. Nachr., Math.-Phys. Kl.* (1914), 177-90.

[24] E. MacCurdy, *The Notebooks of Leonardo da Vinci*, New York, 1941, p. 503. Leonardo also recognized the similarity of air and water (ibid., p. 645).

Prandtl. In modern language, the free steam turbulence induces transition to turbulence in the boundary layer. In turn, this delays separation, thus narrowing the wake and reducing the wake drag.

29. Boundary-layer control

From the beginning,[18] Prandtl thought of his boundary-layer theory as a bridge connecting classical rational hydrodynamics with real hydrodynamics. Theoretically, the boundary layer should tend to become infinitely thin as the Reynolds number R increases without limit. This suggests that by exercising sufficient ingenuity, one should be able to *control* this boundary layer with arbitrarily little effort, so as to approximate to Joukowsky flows having *zero drag and high lift*. The realization of this goal has been the subject of an enormous amount of careful analysis and experimentation, some of which we shall now briefly review.[25]

The most familiar idea involved is that of reducing drag by *streamlining*. By this is meant shaping an airfoil or strut so as to minimize the adverse pressure gradient. This should delay separation, thus reducing drag through the avoidance of leading-edge "stall," or separation near the leading edge. In practice this is achieved by rounding the leading edge and tapering the after-section slowly down to a sharp trailing edge.

The danger of stall, which decreases the lift as it increases the drag, is greatest at high angles of attack. To delay stall, it is also very helpful to curve (or "camber") the wing profile slightly downwards. In the limiting case of a circular arc profile one easily verifies that this device avoids infinite velocity at the leading edge; in general Joukowsky flows it greatly reduces the adverse pressure gradient on the upper (suction) side.

The development of optimum streamline sections for airplanes flying at speeds up to 250 mi/hr (for which compressibility effects are negligible) was a primary occupation of national laboratories in the decades 1910–1930. The analysis of data, and design of new trial sections, was greatly aided by computations of the pressure distributions predicted from Joukowsky theory—and hence from Euler's equations. However, the value of such computations lay *not* in their predictions of lift, drag, or moment (cf. §8), but in their implications as regards transition to turbulence and separation in the boundary layer. How remote from the original concept of Lagrange!

Stall can also be avoided at large angles of attack, and high lift obtained, by the use of so-called *slotted wings*. These are familiar to airplane passengers, as being achieved by wing flaps. Unfortunately, such slotted

---

[25] See [4], Ch. 12; and [23], Ch. 13 for more detailed general introductions to the subject. For Prandtl's ideas see [11], vol. 2, Secs. 50–52 and 91–93.

wings increase the drag; hence they are only used during take-off and landing, when high lift at reduced speeds is the prime consideration. Though the performance of slotted wings can hardly be predicted mathematically, the effect of slots on the flow along the upper sides is evidently similar to that of jets, which reduce the tendency to separation by accelerating the boundary layer. Such jets have also been tried by inventive engineers.

Another promising device for avoiding separation consists in using forced *suction*, either through slots or continuously distributed pores, in the region where the boundary layer would otherwise have separated. This pulls the boundary layer back to the wall, thus again giving a better approximation to Joukowsky flow. If slots are used, one should use Joukowsky theory to concentrate the pressure rise just ahead of the slot.[26] One can also try to use suction to keep the boundary layer *laminar*, thus again reducing drag. Unfortunately, this laminar flow seems very hard to achieve. Dead bodies of flying insects may induce turbulence on the smoothest wing.

A final idea of Prandtl consisted in using *moving boundaries* to prevent boundary-layer retardation and hence separation. Though interesting laboratory demonstrations can be made of the validity of the principle involved,[27] its utilization has been so far on an empirical basis, and so we shall not discuss it further here.

The technical complications involved in developing successfully the devices mentioned above should not obscure the underlying idea: that of approaching ideal Joukowsky flow, as described in §8.

30. Stokes Paradox

In §§25–29 we have been discussing difficulties associated with the theoretical prediction of flows at large R. We shall now pass to the opposite limit, as $R \downarrow 0$. In this limit, expansion in powers of R no longer involves a "singular perturbation" in the sense of §24; the nonlinear convection term $u \cdot \nabla u$ is not of highest order, and it seems mathematically reasonable simply to drop this term.

This was done by Stokes, who thus defined a new class of ideal flows now commonly called "creeping flows." Stokes deduced from this approximation the formula

$$(15) \qquad\qquad D = 6\pi\mu cv$$

[26] See S. Goldstein, *J. Aer. Sci.*, *15* (1948), 189–220; W. Pfenninger, ibid., *16* (1949), 227–36; A. E. von Doehnhoff and L. K. Lofton, Jr., ibid. 729–40; G. V. Lachmann, *J. Roy. Aer. Soc.*, *59* (1955), 163–98.

[27] See [11], Secs. 51–52; J. Ackeret, *Das Rotorschiff* ... , Goettingen, 1925; A. Favre, *Comptes Rendus*, *202* (1936), 434–6. For Prandtl's ideas, see also §9.

for the resistance encountered by a solid sphere of radius c, moving slowly with speed v through a fluid of viscosity μ. The deduction makes essential use of Hypothesis (C), as applied to axial symmetry. The final formula (15) is closely confirmed experimentally[28] for $R < 0.2$.

It would seem only reasonable to apply the same method to circular cylinders moving broadside. However, in this case one has the

Stokes Paradox. No steady "creeping flow" past a fixed circular cylinder is possible.

Proof. In steady creeping plane flow, the Navier-Stokes equations (11) reduce to $\nabla^2 \zeta = 0$. If V is the stream function, this is equivalent to $\nabla^4 V = 0$: V is *biharmonic*. This implies that V is *analytic*;[29] in fact, in any circular annulus one can expand V in Fourier series

$$(16) \qquad V = \sum a_n(r) \cos n\theta + b_n(r) \sin n\theta.$$

where $a_n(r)$ and $b_n(r)$ satisfy

$$(16') \qquad E_n{}^4 a_n = E_n{}^4 b_n = 0, \quad E_n{}^2 = \frac{d^2}{dr^2} + \frac{r^{-1} d}{dr} - n^2 r^2.$$

If the vector velocity, $(\partial V/\partial y, -\partial V/\partial x)$ in rectangular coordinates, is bounded at infinity, then $a_1(r) = A_1 r + a_1/r$, and $a_n(r) = a'_n/r^{n-2} + a''_n/r^n$ if $n \geq 2$, as an easy calculation shows. Hence

$$(17) \qquad V = vy + V_0 + a_0 \ln r + \frac{a_1 \cos \theta + b_1 \sin \theta}{r}$$

$$+ \sum_2^\infty \left\{ \left(\frac{a'_n}{r^{n-2}} + \frac{a''_n}{r^n} \right) \cos n\theta + \left(\frac{b'_n}{r^{n-2}} + \frac{b''_n}{r^n} \right) \sin n\theta \right\}.$$

On the other hand, if $\nabla^4 V = 0$, the Divergence Theorem shows that

$$\iint_A (\nabla^2 V)^2 \, dx dy = \int_C \left\{ \nabla^2 V \frac{\partial V}{\partial n} - V \frac{\partial}{\partial n} (\nabla^2 V) \right\} ds$$

where C is the curve bounding the domain A.

Now let A be the domain between the fixed cylinder and a large circle of radius r. Since $V = \partial V/\partial n = 0$ on the fixed cylinder, the integral on the right side of the preceding equation must vanish over this part of C. On the large circle, since $\nabla^2 V = 0(r^{-2})$, $V = 0(r)$, $\nabla V = 0(1)$, and $\nabla(\nabla^2 V) = 0(r^{-3})$, we have

$$\int_C \left\{ \nabla^2 V \frac{\partial V}{\partial n} - V \frac{\partial}{\partial n} (\nabla^2 V) \right\} ds = 0(r^{-2}) 0(r) \to 0,$$

as $r \to \infty$. Since $(\nabla^2 V)^2 \geq 0$, it follows that $\nabla^2 V \equiv 0$: V must be harmonic.

---

[28] [4], Sec. 215. For (15), see [7], Secs. 337–8.
[29] For the properties of biharmonic functions used, see M. Nicolescu, *Les Fonctions Polyharmoniques*, Hermann, 1936, esp. pp. 13–16 and p. 32. Most classical discussions replace this rigorous analysis by an appeal to Hypothesis (E).

Hence $a'_n = b'_n = 0$ in (17). Finally, by the condition $\partial V/\partial r = 0$ of no slip on the fixed cylinder, $V = V_0$ and $v = 0$, completing the proof.

31. Oseen's equations

Oseen[30] and Lamb have rationalized Stokes' Paradox, pointing out that the convection terms dominate the viscosity term for very large r, no matter how small R is. Overdeterminism can be avoided by a more careful passage to the double limit $R \downarrow 0$, $r \uparrow +\infty$.

To obtain a tractable boundary-value problem, Oseen proposed that one should include linearized convection terms $\sum u_k(\infty)\, \partial/\partial x_k$ in the operator D/Dt, rather than the exact terms $\sum u_k\, \partial/\partial x_k$. By including these terms in Stokes equations, Oseen was able to calculate a theoretical drag formula for a slowly moving cylinder. Approximate experimental confirmation is possible, though difficult ([4], p. 418).

This resolution of the Stokes' Paradox is itself subject to another paradox, discovered by Filon.[31] Filon's Paradox asserts that Oseen's equations, taken literally, would predict an *infinite moment* on an elliptic cylinder obliquely tilted in a stream. This paradox has only recently been resolved by Imai, by going to higher approximations.

Oseen's approximate equations can also be used to correct formula (15) for the influence of small but finite R on the drag of a sphere; the correction factor is $(1 + 3R/8)$. This correction factor has been studied intensively by Goldstein,[32] who has developed a power series for $C_D(R)$ which seems to converge for $R < 2$. Experimental measurements seem to give a lower drag; moreover, in view of the asymptotic nature of Oseen's analysis, one wonders whether the final formula is more than asymptotically valid.[32a]

Solutions of other boundary-value problems involving Oseen's differential equations are reviewed in [9], Part III. However, the approximation to convection forces is quite inaccurate near obstacles and walls. As a result, Prandtl's boundary-layer theory is much more useful at higher Reynolds numbers.

32. Bubble paradoxes

Stokes' Paradox has a close analogue in the theory of underwater explosions. Though there is a simple and extremely useful theory of spherical bubbles due to underwater explosions,[33] it is easy to show that

[30] [9], p. 162; [7], p. 614; see also [12], vol. vi, pp. 29–40.

[31] L. Filon, *Proc. Roy. Soc.*, A113 (1926), 7–27. For Imai's resolution of Filon's Paradox, see ibid., A208 (1951), 487–516.

[32] S. Goldstein, *Proc. Roy. Soc.*, A123 (1929), 225–35; or [4], p. 492. Cf. J. Weyssenhoff, *Annalen der Physik*, 62 (1920), 1–45.

[32a] See S. Kaplun and P. A. Lagerstrom, *J. Math. Mech.*, 3 (1957), 585–93; I. Proudman and J. R. A. Pearson, *J. Fluid Mech.*, 2 (1957), 237–62.

[33] See [7], p. 122; or [15], Ch. XI, §§1–3.

in two-dimensional hydrodynamics any bubble expansion or contraction in an incompressible fluid requires infinite kinetic energy,

$$\tfrac{1}{2} \iint (\rho \nabla U \cdot \nabla U) \, dxdy.$$

With finite forces, compressibility must always be dominant at sufficiently large distances.

There are two other curious bubble paradoxes, whose origin is physical rather than mathematical—and which illustrate the fallibility of Hypothesis (A) of §1. Let a small air bubble rise under buoyancy in a liquid, the bubble being so small that surface tension keeps it nearly spherical and "creeping motion" occurs. Since the bubble is fluid, one must replace (6) by its logical generalization: continuous $u(x)$ across the boundary. The mathematical problem so defined was solved, under the further assumption of continuous tangential stress, by Rybczynsky and Hadamard.[34] The theoretical drag is

$$(18) \qquad D = 6\pi c v \mu \left(\frac{2\mu + 3\mu'}{3\mu + 3\mu'} \right) \simeq 4\pi c \mu v, \quad \text{if} \quad \mu' \ll \mu.$$

In (18), μ is the viscosity of the liquid and μ' is that of the rising bubble.

So much for theory. Experimentally, the drag seems usually to be given by (15) instead of (18). That is, the bubble seems to behave as if it were rigid. This contradiction between theory and experiment may be called the Rising Bubble Paradox.

As suggested by Bond[35] and others, the apparent rigidity may be due to a monomolecular layer of impurities on the bubble interface, which is resistant to deformation.[36] However, the complete story is still not altogether clear.

Even more spectacular is the

Falling Bubble Paradox. In a vertical temperature gradient, thermal variations in surface tension may cause a bubble to fall instead of rise.[37]

Contraction of the bubble "skin" towards the side having greater surface tension, in a viscous liquid, propels the bubble in the direction of decreasing surface tension—i.e. of increasing temperature. This phenomenon seems paradoxical only because it is so unfamiliar, and because surface tension (like viscosity) is almost always conventionally assumed to be a constant in fluid mechanics.

[34] W. Rybczynski, *Bull. Acad. Sci. Cracovie* (1911), p. 40; J. Hadamard, *Comptes Rendus*, *152* (1911), p. 1735. For the experimental data, see G. Barr, *A Manual of Viscometry*, Oxford, 1931, p. 190 ff.; T. Bryn, *Forschung Ing.*, *4* (1933), 27–30; [11], vol. 2, Sec. 74.

[35] W. N. Bond, *Phil. Mag.*, *4* (1927), 889–98. For an excellent review of the subject see J. S. McNown, *La Houille Blanche*, *6* (1951), 701–22.

[36] The resistance may be viscous or elastic; see D. W. Criddle and A. L. Meader, Jr., *J. Appl. Phys.*, *26* (1955), 838–42, and the refs. given there.

[37] M. J. Block, N. O. Young, and J. S. Goldstein, *J. Fluid Mech. 6* (1959), 350–6. See [15], p. 319.

33. Bulk viscosity

As indicated in §19, the usual (metaphysical) derivation of the Navier-Stokes equations (1*) involves two coefficients of viscosity, λ and μ. One can suppose the coefficient μ of shear viscosity to have been measured in Poiseuille flow; this leaves the problem of measuring λ and verifying the consequences of (1*) for this λ—which is presumably a function of temperature T and pressure p.

As stated in §19, Stokes simply assumed that $\lambda = -2\mu/3$. However, it is clearly safer to define $\mu' = \lambda + 2\mu/3$ and to study it experimentally. Physically, λ and the bulk viscosity μ' are meaningless until they have been defined and measured experimentally. This is not easy. Thus, in compressible boundary layers the magnitude of μ' is not too important because the rate of shear deformation so greatly exceeds the rate of compression. Partly for this reason, bulk viscosity is commonly neglected in high-speed flow theory,[38] and most careful authors are non-committal about the Poisson-Stokes relation $\mu' = 0$.

Experimental determinations of μ' are usually based on measurements of sound attenuation, but the proper interpretation of such measurements is far from simple. Thus, Stokes' own theory of sound attenuation not only assumed $\mu' = 0$, attributing all attenuation to μ; it ignored dissipation by heat conduction (thermal diffusion) as well. Kirchhoff repaired the latter omission and also calculated the (often much larger) attenuation due to boundary-layer friction for sound propagated in tubes;[39] the latter does involve only μ. But neither author seems to have viewed observations of sound attenuation as a means for measuring μ'.

Though the proper interpretation of the experimental data is still somewhat controversial, the following facts seem fairly clear. In some gases, such as He, A, and N_2, the observations[40] seem compatible with the assumption $\mu' = 0$. But other gases such as O_2 and CO_2 give much more rapid attenuation in specific *frequency bands*.[41] In air, disproportionately large effects may be caused by a small relative humidity or CO_2 content, as well as by dust or (in tubes) wall roughness. In most liquids, absorption is highly dependent on frequency; moreover, the bubble content must be

[38] See [5], pp. 36–8; and [10], Ch. 11. For remarks about the assumption $\mu' = 0$, see [7], Secs. 325, 328; [10], p. 185; and [11], vol. 1, p. 260, ftnt.

[39] See [7], Secs. 359–60; G. Kirchhoff, *Pogg. Ann.*, *177* (1868), 177–93. For boundary-layer friction, cf. §115.

[40] This may be related to Maxwell's proof (by molecular considerations, see §34) that $3\lambda + 2\mu = 0$ in an ideal monatomic gas; see also Y. Rocard, *Hydrodynamique et Théorie Cinétique des Gaz*, Appendix I.

[41] A. W. Duff, *Phys. Rev.*, *6* (1898), 129–39, and *11* (1900), 65–74; A. van Itterbeck and P. Mariens, *Physica*, *4* (1937), 207–15 and 609–16; J. J. Markham, R. T. Beyer, and R. B. Lindsay, *Revs. Mod. Phys.*, *23* (1951), 353–11; R. S. Marvin et al., *J. Appl. Phys.*, *25* (1954), 1213–18.

carefully watched. Thus a 0.17% relative bubble volume will reduce the speed of the speed of sound by a factor of five,[42] and augment sound attenuation accordingly.

It is easiest to explain the dependence of (ultrasonic) sound attenuation on frequency by postulating *relaxation times*[43] (hysteresis effects) for the transfer of molecular energy between different modes. However, various authors have tried to reconcile the observational data with suitable interpretations of bulk viscosity[44]—i.e. of λ in (1*).

34. Molecular effects

Molecular effects explain many phenomena in real fluid dynamics, besides the frequency-bands for ultrasonic sound absorption described in §33. It would be forced and unnatural to treat most of these phenomena within the orthodox framework of continuum mechanics.

For example, one obviously needs to consider relaxation times in estimating the thickness of shock fronts,[43] in fluids where the bulk viscosity is influenced by such molecular hysteresis effects. (In the classical theory of continuum mechanics the shock front thickness is assumed to be zero.) In the upper atmosphere, where the density (pressure) is very low, relaxation times also have an important effect on the distance of *shock detachment*[44a] for $M > 2$.

Again, molecular *dissociation*, recombination, and ionization affect the shock thickness in hypersonic flow;[44b]; in fact, they have a considerable effect on the entire fluid motion when the Mach number $M > 10$ under atmospheric conditions. Thus, air contains 1% NO at 2,000°K, and 10% NO at 3,000°K. Above 11,000°K, ionization is appreciable.

For this reason realistic studies of hypersonic shocks must take increasing account of chemical physics.[45]

The orthodox framework of continuum mechanics is also unable to

[42] A. B. Wood, *A Textbook of Sound*, p. 362. For surprising pressure effects, see T. A. Litovitz and E. H. Carnevale, *J. Appl. Phys.*, *26* (1955), 816–20.

[43] H. O. Kneser, *Annalen der Physik*, *11* (1931), 761–76; C. Eckart, *Revs. Mod. Phys.*, *20* (1948), 232–5; L. N. Liebermann, *Phys. Rev.*, *75* (1949), 1415–22; M. J. Lighthill in [25], Sec. 4 (for shock thicknesses, see Sec. 6); K. F. Herzfeld and T. A. Litovitz, *Absorption and Dispersion of Ultrasonic Waves*, Academic Press, 1959.

[44] L. Tisza, *Phys. Rev.*, *61* (1942), 531–6; S. B. Gurevitch, *Doklady USSR*, *55* (1947), 17–19; S. Karim and L. Rosenhead, *Revs. Mod. Phys.*, *24* (1952), 108–16; C. Truesdell, *J. Rat. Mech. Analysis*, *4* (1953), 643–721.

[44a] See, for example, R. N. Schwartz and J. Eckerman, *J. Appl. Phys.*, *27* (1956), 169–74. If the effect were one of instantaneous dissociation, the distance would be density-independent (§73).

[44b] See M. J. Lighthill, *J. Fluid Mech.*, *2* (1957), 1–32; also [4a].

[45] See *Nature*, *178* (1956), 343–5; W. E. Deal, *J. Appl. Phys.*, *28* (1957), 782. A valuable survey of World War II research has been made by W. G. Penney and H. H. M. Pike, *Progress in Physics*, *13* (1950), 46–82.

explain the physical phenomenon of *diffusion* and its reverse, the *separation* of a composite liquid or gas into its constituents—e.g., of cream from milk. In a perfect continuum, isotope separation would be impossible, whether by gaseous diffusion, in a centrifuge, or in a nozzle.[45a]

Slip flow. A quite different application of molecular ideas is to the breakdown of the no-slip condition (6) when the molecular mean free path is comparable to the macroscopic dimensions. Three important special cases are: flow in narrow pores, the free fall of minute drops (Millikan's oildrop experiment), and satellite retardation. In all cases, appreciable departures from the laws of continuum mechanics are involved: [46] observed shear forces are appreciably less than those predicted by (13) and (15).

These effects are most easily interpreted as corresponding to a partial breakdown of (6). In the extreme case of "specular reflection" of all molecules (angle of incidence equals angle of reflection), evidently the shear stress on the boundary would be zero. Thus (6) could logically be replaced by the condition of continuous normal velocity (Ch. I, (7)), appropriate to nonviscous flow. Since $\nabla^2 u_i = \partial(\nabla^2 U)/\partial x_i = 0$ in any irrotational incompressible flow, any solution of the Neumann problem (Euler or Joukowsky flow) would satisfy the Navier-Stokes equations (5). Thus, specular reflection would imply a complete reorientation of the theory of viscous flows.

To interpret the experimental facts, a quasi-empirical "accommodation coefficient" is assumed, instead of specular reflection. This assumption, combined with a consideration of the finite mean free path postulated in the kinetic theory of gases, gives predictions in general agreement with observational data.

"Slip" in rarefied gas flow is not to be confused with the nineteenth-century concept of bulk slip at the boundary of very smooth solids (e.g. Hg against glass). Thus Stokes[47] thought that slip must occur "above a certain velocity," while many other eminent scientists kept their fingers crossed. In view of the many peculiarities of surface physics, this is not too surprising. Also, an assumed analogy with solid friction—in which the shear stress τ is bounded by a constant $\mu < 1$ times the normal pressure— would support Stokes' idea. Even today, though the evidence against bulk slip is almost overwhelming,[48] certainty is not absolute and universal.

[45a] For aerodynamic ("nozzle") methods of isotope separation, see M. Leroy, *Nucleonics*, April, 1960, p. 68.

[46] M. Knudsen, *Annalen der Physik, 34* (1911), 593–656. For recent surveys with references see G. N. Patterson, *Molecular Flow of Gases*, Wiley, 1956, Ch. V; also S. Schaaf, Chs. 9–10 in *Heat transfer*, Univ. of Michigan Press, 1953; [3a], Sec. H.

[47] [13], vol. 1, pp. 96–9. See also p. 186 there.

[48] For a scholarly review of the facts about bulk slip, see [4], pp. 676–80. For recent data see M. R. Brockman, *Nat. Bu. Standards Rep. 4673* (1956); E. Schnell, *J. Appl. Phys.*, *27* (1956), 1149–52; P. Debye and R. L. Cleland, ibid., *30* (1959), 843–9.

35. Conclusions

The evidence briefly reviewed above favors strongly the validity of the Navier-Stokes equations for *incompressible* viscous fluids, as regards the bulk flow of most common gases and liquids at flow speeds much smaller than their speeds of sound (i.e. if $M < 0.2$). However, in the majority of applications one cannot rely on the plausible hypotheses listed in §1, useful though these hypotheses are in other contexts. Because of *turbulence* especially, a very careful use of statistical analysis on a high mathematical level seems to be required.

Thus, instead of viewing the Navier-Stokes equations as defining obvious boundary-value problems (e.g. the steady-flow problem), one must look to the *physical evidence* as regards the proper boundary-value problems to be set.

Though often forgotten, this situation is by no means unfamiliar in analysis. Thus, in his address inaugurating the First International Mathematical Congress, Poincaré said, "However varied may be the imagination of man, nature is still a thousand times richer," and again, "Each of the theories of physics . . . presents (partial differential) equations under a new aspect . . . without these theories, we should not know partial differential equations."[49] These fundamental truths have since been reaffirmed by other eminent mathematicians and should always be kept in mind.

The properties of *compressible* viscous fluids which may legitimately be assumed are more obscure. Though it is legitimate to hope that some variant of the Navier-Stokes equations will survive a critical examination of the physical facts, how such a variant should be combined with thermodynamics is still obscure.

For complete rigor, one must take into account heat conduction and radiation as well as viscous heating and the variation of viscosity and density with temperature. Rigorous solution of the resulting boundary-value problems—and proof that they are mathematically "well-set"—seems like a nearly hopeless task. Equally difficult, if not more so, would be the rigorous application of perturbation methods, justifying the neglect of selected variables. Progress, if it is to be made at all, will have to depend on Hypotheses (A)–(F) of §1, and other similar heuristic assumptions.

However, enough for generalities! We shall now make case studies of three special concepts of fluid dynamics, so as to get a sharper picture of the kinds of consideration which may be relevant. These are the concepts of free boundary, of similarity, and of induced mass.

[49] H. Poincaré, *Proc. First Int. Math. Congress*, Zürich, 1897. pp. 81–90. For reaffirmations, see J. Hadamard, *Lectures on Cauchy's Problem*, Yale Univ. Press, 1923, p. 23; and R. Courant, *Proc. Eleventh Int. Math. Congress*, Cambridge (U.S.A.), 1950, vol. *2*, p. 278.

III. *Jets, Wakes and Cavities*

36. Discontinuous flows

In real fluids of small viscosity one commonly observes that flows tend to *separate* from rigid walls, especially at sharp corners. This has already been remarked in §8, where Fig. 2c depicts such a flow, and in §29.

Mathematical models of such flows with separation can be constructed quite easily, within the framework of Euler's equations of motion for a *nonviscous* fluid. The essential idea is to let the velocity change discontinuously across streamlines, thus violating brutally Hypothesis (E) of §1. Simple examples of such flows are sketched in Figs. 9a–9b. In these

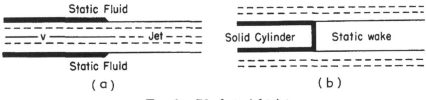

Static Fluid

v ———— Jet ———

Static Fluid

(a)

Solid Cylinder | Static wake

(b)

FIG. 9a. Ideal straight jet

FIG. 9b. Ideal wake behind half-body

flows all streamlines are parallel, and regions of uniform flow are separated from regions of stagnant water by *streamlines of discontinuity*, across which the velocity changes discontinuously. Fig. 9a depicts an idealized infinite *jet* entering still water from a tube of arbitrary cross-section, while Fig. 9b depicts a uniform stream separating from a half-body where the latter is chopped off, and enclosing a stagnant *wake* behind the half-body. In both cases the pressure can be taken as hydrostatic.

By definition, the shear stress is zero in an ideal nonviscous fluid; hence, a necessary and sufficient condition for the dynamic equilibrium of a streamline of discontinuity is the *continuity of pressure* across it. If the streamline bounds an idealized "wake" or other region full of motionless fluid ("dead water"), and gravity is compensated for as in Theorem 1 of §21, then continuity of pressure is tantamount to constancy of pressure in the wake. Hence, by the continuity of pressure, streamlines bounding wakes must be at constant pressure. Streamlines of discontinuity at constant pressure are called *free streamlines*.

By the Bernoulli equation (8*) of Ch. I, still neglecting gravity, the velocity is constant along any free streamline in steady flow, and conversely: $v \, dv = -dp/\rho = 0$. This gives a purely kinematic boundary condition for steady flows bounded by free streamlines. Combined with the results of §5, it defines the following boundary-value problem of potential theory.

Helmholtz Problem. Given an obstacle R, find a velocity potential which satisfies (i) $\nabla^2 U = 0$ outside R and a "dead water" region R_1, (ii) $\partial U/\partial n = 0$ on the boundaries of R and R_1, and (iii) $|\nabla U|^2 = \text{const.}$ on the boundary of R_1.

Note that the last boundary condition is *nonlinear*. Note also that the flow topology has been left unspecified; in practice it is postulated on intuitive or experimental grounds (Hypothesis (D) of §1). Two such flow

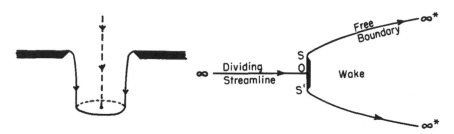

FIG. 10a. Circular jet   FIG. 10b. Wake behind disc

topologies are sketched in Figs. 10a–10b. They represent, respectively, the "jet" from a circular hole in a plane wall, and the "wake" behind a disc.

Flows satisfying conditions (i)–(iii) above—i.e. solutions of the Helmholtz problem,—will be called *Helmholtz flows* in the sequel.

Actually, nobody has yet succeeded in giving exact mathematical treatments of these two Helmholtz flows; see §49. However, their plane analogs—i.e. the jet from a slot and the wake behind a flat plate—can be constructed quite easily. The construction of these plane Helmholtz flows will be the object of §§37–39.

37. Semicircular hodographs

In general, locally irrotational, incompressible plane flows are characterized by the existence of *complex potentials* $W = U + iV$. Here U is the velocity potential and V is the stream function. The complex potential W is an analytic function of the complex position variable $z = x + iy$, and the derivative

$$(1) \qquad \frac{dW}{dz} = \zeta = u - iv$$

is the *conjugate* of the complex velocity $u+iv=\zeta^*$

$$\left(u=\frac{dx}{dt},\quad v=\frac{dy}{dt}\right).$$

If one knows $W=f(\zeta)$ as a complex analytic function of ζ, therefore, one can determine z as the analytic function

(2) $$z=\int \zeta^{-1}f'(\zeta)\,d\zeta=\int \zeta^{-1}\,dW$$

of ζ also, and hence (in principle) eliminate ζ to find $W(z)$. Hence, to determine a steady plane Helmholtz flow, it suffices to know the functional relation $W=f(\zeta)$. This can be guessed, in the plane analogs of Figs. 10a–10b, by considering the hodographs of the two flows in question.

By the *hodograph* of a plane flow is meant the locus of those values of ζ actually attained in the flow. From Figs. 10a–10b one easily guesses that the hodographs of the associated plane flows (if they exist) should be *semicircular*. For, arg ζ (the flow direction) is constant on the flat plates (fixed boundaries), while $|\zeta|$ (the flow speed) is constant along the free boundaries, as explained in §36.

On the other hand, the W-*domain*, or locus of values of W attained in the flow, is bounded by lines $V=$ const. (streamlines) parallel to the real U-axis. In Fig. 10a, it is an infinite strip; in the wake of Fig. 10b it is a half-plane cut along the positive U-axis (choosing the constant of integration of $W=\int \zeta\,dz$ so as to make $W=0$ at the stagnation point).

Conformal mapping technique. Now let Φ be any flow having a semicircular hodograph (hence bounded by free streamlines and straight walls). We can choose axes making ζ real on the fixed boundary, and velocity units making $|\zeta|=1$ on the free boundary. Then, map the hodograph onto the lower half-plane Im $\{\sigma\}<0$ by the transformation $\sigma=(\zeta+\zeta^{-1})/2$. The most general conformal transformation mapping the hodograph onto the lower half-plane is then

(3) $$T=\frac{a\sigma+b}{c\sigma+d}=\frac{a(\zeta^2+1)+2b\zeta}{c(\zeta^2+1)+2d\zeta},\quad ad>bc,$$

with real coefficients a,b,c,d.

On the other hand, the W-domain of any simply-connected flow bounded by streamlines is a generalized "polygon," with one or more vertices at infinity and all sides parallel to the real axis. Hence, the T-domain (3) can be mapped onto the W-domain by a suitable (conformal) *Schwarz-Christoffel* transformation

(4) $$\frac{dW}{dT}=R(T)=\frac{C\prod_j (T-B_j)}{\prod_k (T-T_k)},$$

where C and the B_j, T_k are *real* parameters ([6], p. 370). By (3) and (4) we see that for any simply-connected flow Φ with semicircular hodograph, we can express $W = \int A(\zeta)\, d\zeta$, where $A(\zeta)$ is a *rational function*.

The preceding technique can be generalized easily to the case of a circular sector hodograph, subtending an angle π/n ([15], Ch. II). For, in this case the transformation ζ^n maps the hodograph onto a semicircle, and hence

$$T = \frac{a(\zeta^{2n}+1)+2b\zeta^n}{c(\zeta^{2n}+1)+2d\zeta^n}$$

maps it onto a half-plane. From here we can proceed as before.

38. Jet from slot

Helmholtz [19] first applied the preceding technique in 1868, to the case of a plane jet from a slot sketched in Fig. 11a. In this case, choosing

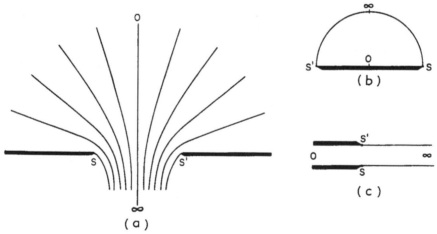

FIG. 11. Plane jet from a slot

the unit of length so as to make V jump by π across the orifice, we can set $W = \ln T$, $T = e^W$, $dW/dT = 1/T$ in the formulas of §37.

The choice of a, b, c, d is determined by considering the correspondence between ζ and W in the physical plane. The limit $\zeta = 0$ in the hodograph of Fig. 11b is evidently approached as $W = -\infty$ in the W-domain of Fig. 11c, or equivalently, as $T = 0$. This makes $a = 0$ in (3). Up to similarity, we can therefore write

(5) $T = \dfrac{\zeta}{\zeta^2 - 2C\zeta + 1}$, $W = \ln T$, $C = -d/c$.

The point at infinity on the jet, where $W = +\infty$, evidently corresponds to $\zeta = e^{i\alpha}$, $(\zeta + \zeta^{-1})/2 = \sigma = C = \cos \alpha$. Hence Eqs. (2) and (5) define a *jet from a slot* in an infinite plate, the jet making an angle α with the plate.

Since the integrand is a rational function, we can integrate (2) in closed form, using (5), getting

$$(6) \qquad W = \ln \zeta - \ln(\zeta^2 - 2C\zeta + 1);$$

we omit the detailed computations. The case of a *vertical* jet considered by Helmholtz is just the case $C = \cos \alpha = 0$.

39. Kirchhoff wake

In 1869 Kirchhoff [33] carried out an analogous treatment for the wake behind a plate. In this case $W = T^2$ maps the lower half-plane onto the cut plane constituting the W-domain; thus $R(T) = 2T$ in (4), *provided* the real axis is made to run along the plate.

To determine the constants a, b, c, d in (3), we again consider the correspondence between W and ζ in the physical plane. The point $W = 0$ where the cut begins occurs at the stagnation point of the flow, where $\zeta = 0$. Hence $a = 0$ and we can again write (5), with the understanding that $C = \cos \alpha$, the flow direction at infinity.

The case of *normal* incidence is the most interesting; it describes the plane analog of Fig. 10b. In this case (5) reduces to

$$(7) \qquad W = T^2 = \left(\frac{\zeta}{\zeta^2 + 1}\right)^2$$

Performing the integration indicated symbolically in (2), we get this time

$$(8) \qquad z = \frac{\zeta}{(\zeta^2 + 1)^2} + \frac{1}{2}\left\{\frac{\zeta}{\zeta^2 + 1} - \frac{i}{2} \ln \frac{\zeta - i}{\zeta + i}\right\} + \text{const.},$$

for all values of $\zeta = \xi + i\eta$.

Along the plate, ζ is real and (8) reduces to

$$(8a) \qquad z = \frac{\zeta}{(\zeta^2 + 1)^2} + \frac{1}{2}\left\{\frac{\zeta}{\zeta^2 + 1} + \arctan \zeta\right\},$$

where clearly $z(0) = 0$, and so the constant of integration vanishes. At the right-hand separation point S, $\zeta = 1$ and so

$$(8b) \qquad z(S) = \frac{1}{4} + \frac{1}{2}\left\{\frac{1}{2} + \frac{\pi}{4}\right\} = \frac{4 + \pi}{8}.$$

The pressure is most easily computed by setting $\zeta = \xi = \tan \theta$ along the plate, so that $\zeta/(\zeta^2+1)=\sin\theta\cos\theta=1/2\sin 2\theta$. Hence the thrust on half the plate is, by Bernoulli's Theorem and (8a),

$$\int_{\xi=0}^1 (1-\xi^2)\,dx = \int_{\xi=0}^1 \frac{1-\tan^2\theta}{\tan\theta}\,d[\tfrac{1}{4}\sin^2 2\theta]$$

$$= \int_{\theta=0}^{\pi/4} 2\cos^2 2\theta\,d\theta = \frac{\pi}{4}.$$

The ratio of half the thrust to the plate half-length is clearly the *drag coefficient*, which is thus

(9) $$C_D = \frac{2\pi}{\pi+4} = 0.88 \quad \text{approximately.}$$

Similar but more elaborate computations give, for the case of *oblique* incidence $\alpha\neq\pi/2$,

(9') $$C_D = \frac{2\pi\sin^2\alpha}{4+\pi\sin\alpha}, \qquad C_L = \frac{\pi\sin 2\alpha}{4+\pi\sin\alpha}.$$

40. Wall effects

The hodograph method[1] can also be applied to give useful information about wall effects in jets from nozzles.

Consider the Helmholtz flow past a plate of half-width b, held symmetrically in the jet from a nozzle, as in Fig. 12a. Much as before, the functions $W(z)$ and $\zeta(z)$ map the flow conformally onto a slit infinite strip and a semicircle, respectively. This is again because the W-domain is bounded by streamlines, on which $V=$ const, including the dividing streamline which branches at the stagnation point—while the hodograph is bounded by the free streamline, on which the magnitude of ζ is constant, and the fixed plate, on which the direction of ζ is vertical.

Mathematically, it is convenient to locate the origin and to choose units so that the W-domain occupies the strip $-\pi<V<\pi$, slit along the positive semi-axis $W=U>0$, and so that the hodograph is the unit semicircle $|\zeta|<1$, $-\pi/2<\arg\zeta<\pi/2$. It is also convenient to consider only the lower *half* of the flow.

If this is done, then $T=e^W$ occupies a half-plane. Since ζ^2 maps the half-hodograph onto the unit semicircle, $0<\arg\zeta<\pi$, $|\zeta|<1$, $(\zeta^2+\zeta^{-2})/2$ maps it onto a half-plane. By the Fundamental Uniqueness Theorem on Conformal Mapping, it follows that

(10) $$W = \ln T, \qquad T = \frac{a(\zeta^2+\zeta^{-2})+2b}{c(\zeta^2+\zeta^{-2})+2d}, \quad a, b, c, d \text{ real.}$$

---

[1] See [15], Ch. II, §§7–8; also Ch. I, §11.

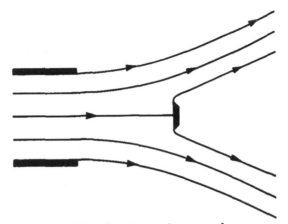

FIG. 12a. Plate in jet from nozzle

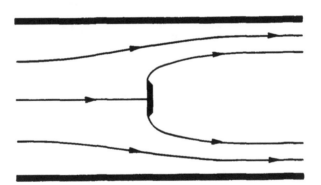

FIG. 12b. Plate in channel

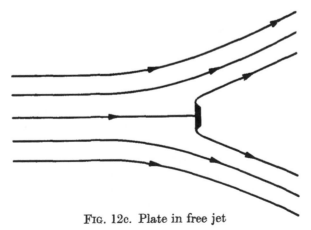

FIG. 12c. Plate in free jet

The real constants a, b, c, d in (10) can be correlated with geometrical features of the flow. If $v < 1$ is the velocity in the nozzle (assumed constant), and $e^{-i\alpha}$ that in the lower jet, then $\zeta = v$ when $W = -\infty$ and $T = e^W = 0$, while $\zeta = e^{i\alpha}$ when $W = T = \infty$. Thus, neglecting additive constants, we have

$$
(11) \qquad
\begin{aligned}
W &= \ln[(\zeta^2 - e^{2i\alpha})(\zeta^2 - e^{-2i\alpha})] - \ln[(\zeta^2 - v^2)(\zeta^2 - v^{-2})] \\
&= \ln[\zeta^4 - 2C\zeta^2 + 1] - \ln[\zeta^4 - (v^2 + v^{-2})\zeta^2 + 1],
\end{aligned}
$$

where $C = \cos 2\alpha$. Since $z = \int \zeta^{-1} \, dW$, by (2), this can be integrated by elementary methods to give $z(\zeta)$ in closed form.

Of especial interest are the cases $C = 1$ of a plate in a *closed channel* (Fig. 12b), in which case

$$
(12a) \qquad z = 4 \tanh^{-1} \zeta - 2v \tanh^{-1} v\zeta - \frac{2}{v} \tanh^{-1} \frac{\zeta}{v},
$$

and the case $v = 1$ of a plate in a *free jet* (Fig. 12c) of width π, when

$$
(12b) \qquad z = 2 \ln \frac{1+\zeta}{1-\zeta} - e^{i\alpha} \ln \frac{1 + e^{i\alpha}\zeta}{1 - e^{i\alpha}\zeta} - e^{-i\alpha} \ln \frac{1 + e^{-i\alpha}\zeta}{1 - e^{-i\alpha}\zeta}.
$$

Using (12a)—(12b), one can show that the wall correction for C_D is small if based on the free streamline velocity, but very large if based on the upstream velocity. Thus, it is 30% if the (two-dimensional) tunnel is 100 diameters wide. In a free jet the correction is small and the above problem does not arise.

41. Helmholtz instability

Unfortunately, the free boundaries of the jets and wakes considered by Helmholtz and Kirchhoff are *unstable*. This was already known by Helmholtz ([19], p. 222), who observed qualitatively that the boundaries of jets from organ pipes "rolled up" into periodically spaced spirals.

Observation also shows that at Reynolds numbers $R > 10^4$ the streamlines which separate from a flat plate (or other obstacle) in a moving stream soon break up into a turbulent "mixing zone." As a consequence, real wakes are by no means motionless bands of "dead water" extending to infinity, as postulated by Kirchhoff. Real wakes are filled with eddies, most active in the mixing zone; these continually carry fluid downstream from the wake periphery.

Maintenance of the fluid content of the wake requires that a mean counterflow should exist in the center of the wake. This results in two eddies in the mean flow, as sketched in Fig. 13. These are maintained by a considerable pressure gradient; they make the wake pressure p_w just behind the plate considerably less.

Related to this extreme instability, there is a large underpressure in real wakes: the wake pressure p_w is considerably less than the ambient pressure p_a. As a consequence, the Kirchhoff prediction of $C_D = 0.88$ is less than half of the real C_D, which is about two. For tilted plates the lift C_L is even more seriously underestimated by the Kirchhoff model, especially for angles below the "angle of stall" (about 15°).

Were it not for this fact, airplane flight would be extremely difficult. This was observed by Rayleigh ([12], vol. i, p. 287, and vol. iii, p. 491) as

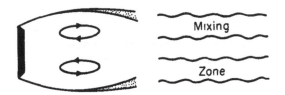

FIG. 13. Counterflow in real wake

early as 1876. Fortunately, the Joukowsky model of §8 gives a much better approximation to reality at small angles of attack. (Moreover, flow separation can be greatly delayed by proper airfoil design, as already explained in §29.)

These facts were all well known to Kelvin.[2] Kelvin went much farther: he showed how to analyze quantitatively the stability of straight stream-lines in parallel flow.

Of greatest interest is the case of a horizontal interface in a vertical

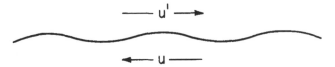

FIG. 14. Helmholtz-Kelvin instability

gravity field, separating two fluids of densities ρ, ρ' moving with velocities u, u', as in Fig. 14. Kelvin showed that under these circumstances a sinusoidal perturbation of the interface with wave-length $\lambda = 2\pi/k$ should grow like $\exp\{I(\lambda)t\}$, where

$$(13) \qquad [I(\lambda)]^2 = \frac{\rho\rho'k^2}{(\rho+\rho')^2}(u'-u)^2 - \frac{\rho-\rho'}{\rho+\rho'}gk - \frac{\gamma k^3}{\rho+\rho'},$$

γ being the interfacial tension, and g, the gravitational acceleration. His argument is reproduced in [7], p. 462.

In the case $\rho = \rho'$, $\gamma = 0$ of wakes, clearly $I(\lambda) = k|u'-u|/2 > 0$. The

[2] *Nature, 50* (1894), pp. 524, 549, 573; see also Rayleigh [22], vol. 6, p. 39.

interface is highly unstable, and the rate of growth of perturbations of very short wave-length is *unbounded*. In n wave-lengths $|u' - u|t = n\lambda = 2\pi n/k$ of relative motion, the amplification is by a factor $e^{2\pi n}$.

Wind-generation of waves. Kelvin gave a classic application of his formula (13), to the prediction of the minimum wind-speed required to generate surface ripples on still water. It is a matter of common observation that in sufficiently light breezes the surfaces of ponds retain a mirror-like smoothness. Theoretically, for a uniform wind blowing air of standard relative density $\rho'/\rho = .00126$ over water, one can show that formula (13) predicts that disturbances of all wave-lengths will be neutrally stable, if and only if $|u' - u| < 646$ cm/sec. In practice, waves can be generated by winds whose mean velocity is one-fifth this amount; a simple and convincing theory of this paradox has yet to be invented.[3]

42. Cavities

When a solid moves through a liquid at high speed, its "wake" is typically gas-filled. Such gas-filled wakes are called *cavities*, and solutions of the Helmholtz problem describe flows around cavities much better than they describe wakes.

Cavities occur naturally under many conditions. Thus, air-filled cavities may be photographed [38] behind spheres dropped into water from heights of two meters or more. Vapor-filled cavities also form behind underwater missiles whose speeds exceed (say) 30 meters/sec. Such cavities also normally form on the blades of marine propellers at thrust pressures on the propeller disc exceeding about 20 psi, and the collapse of the small bubbles associated with "incipient cavitation" in such cases is a dreaded cause of propeller erosion. Similar erosion may occur, and for the same reason, in overloaded hydraulic turbines. Paradoxically, "supercavitating" propellers having much higher thrust and operating in large cavities can be designed so as to avoid this erosion.

The practical importance of cavitation was first recognized around 1900. Its mathematical analysis is based on the plausible assumption that vapor-filled cavities will form by evaporation as soon as the pressure falls below a well-defined "vapor pressure" p_v. In mathematical form this is equivalent to the condition

$$(14) \qquad p = p_v \quad \text{in the cavity;} \qquad p \geqq p_v \quad \text{in the liquid.}$$

(In air-filled cavities $p > p_v$ is of course possible.)

---

[3] F. Ursell has recently reviewed the problem in [25], pp. 216–49, summarizing some experimental data on p. 240.

Guided by this idea, Thoma introduced[4] in 1924 the now widely used *cavitation parameter* (cavitation number)

$$(15a) \qquad\qquad Q_v = \frac{p_a - p_v}{\frac{1}{2}\rho v^2},$$

p_a being the ambient pressure. Not long afterwards it was shown by Ackeret and others[5] that (14) and the theory of Helmholtz flows were applicable to large-scale cavities behind solid obstacles. But it was primarily the development of underwater missiles for use in World War II which led to the development of cavity theory as known today.

43. Parameters ρ'/ρ and Q

The parameters ρ'/ρ and Q_v defined empirically above, though mentioned occasionally in the engineering literature, were essentially ignored in textbooks on rational hydrodynamics[6] prior to 1945. Today, much of the study of Helmholtz flows is keyed to them. For example, a consideration of ρ'/ρ explains why Helmholtz flows represent steady cavity flows and liquid jets in air (i.e. *two-phase* flows) so much better than they represent wakes or (say) gas jets.

In the case of high-speed missiles, the terms in g and γ are relatively small in (13). Hence, if $\rho' \ll \rho$, we have $I(\lambda) = (k|u' - u|/2)\sqrt{\rho'/\rho}$, to a first approximation. This leads one to expect the *distance* $|u' - u|t$ from the separation point to the mixing zone, where the Helmholtz model breaks down for a given wave-length λ, to be proportional to $\sqrt{\rho/\rho'}$. With air-filled cavities, $\rho/\rho' \simeq 750$; with vapor-filled cavities, $\rho/\rho' \simeq 30{,}000$; hence in both cases Kelvin's analysis leads one to expect the free streamline instability to be small.

This explains rationally the empirical dictum of Betz and Petersohn,[7] based on Ackeret's work and earlier verifications by von Mises of theoretical predictions for jets of water into air, that *free streamline theory is applicable if* $\rho'/\rho \ll 1$. For example, though the wall effects deduced in §40 do not apply to real wakes, for which they were first derived,[8] they are very important for real cavities.

[4] D. Thoma, *Trans. First World Power Conf.* (1924), vol. *2*, 536–51; see also H. B. Taylor and L. F. Moody, *Mech. Engineering*, *44* (1922), 633–40. See §72.

[5] J. Ackeret, *Tech. Mech. und Thermodynamik*, *1* (1930), 1–21 and 63–72. In 1932, Weinig first applied the Riabouchinsky model of §7 to cavity flows.

[6] See [7], Secs. 73–80 and Ch. XII of the first edition of [8]. The contrast with [14], Ch. II, is quite striking; see also *Proc. Seventh Int. Congress Appl. Mech.*, London, 1948, vol. *2*, pp. 7–16.

[7] *Ingenieur Archiv*, *2* (1931), 190–211. For von Mises' data, which confirm the formulas derived in §40, see *Zeits. VDI*, *61* (1917), 447–52, 467–73, and 493–7.

[8] V. Valcovici, Inaugural dissertation, Goettingen, 1913. For applications to cavities, see G. Birkhoff, M. Plesset, and N. Simmons, *Quar. Appl. Math.*, *8* (1950), 151–68; and *9* (1952), 413–21.

Practical applications of free streamline theory depend also on a second parameter, which would coincide with (15a) if (14) were exact. Assuming (14) and the Bernoulli equation in ideal two-phase Helmholtz flow, (15a) becomes $Q_c = (v_f/v_a)^2 - 1$, where v_f is the free streamline velocity, and v_a, the ambient ("free stream") velocity. Hence, we define the *ideal* cavitation parameter in a Helmholtz flow as

$$(15b) \qquad\qquad Q = \left(\frac{v_f}{v_a}\right)^2 - 1.$$

Clearly, in an ideal fluid (14) implies $Q > 0$. In air-filled cavities also, the empirical cavity underpressure coefficient

$$(15c) \qquad\qquad Q_c = \frac{p_a - p_c}{\frac{1}{2}\rho v^2}, \quad p_c = \text{cavity pressure},$$

is always positive. Finally, the *wake underpressure* $(p_a - p_w)$, mentioned in §41, expressed dimensionlessly in terms of the wake underpressure coefficient

$$(15d) \qquad\qquad Q_w = \frac{p_a - p_w}{\frac{1}{2}\rho v^2}.$$

is empirically between zero and one. Thus, behind a flat plate Q_w is about one.

By accepting Q_w as an empirical parameter, one can even apply free streamline theory usefully to *wakes*. Thus, if one introduces a numerical correction factor $(1 + Q_w)$, so as to match the observed wake underpressure just behind an inclined flat plate, good agreement is obtained between the formulas of the Kirchhoff theory and observed pressure distribution functions on the front side ([15], p. 28, Fig. 3)—at least if $\alpha > 15°$, above the "angle of stall."

The condition $Q > 0$ can clearly be identified, in nonviscous flow, with the following purely kinematic condition, introduced in 1911 by M. Brillouin [16] in connection with wakes.[9]

Brillouin Condition. The velocity is maximized along the free streamline.

Though $Q > 0$ in all practical applications known to me, it is incorrect to assume that (14) holds rigorously under all circumstances. See [15], Ch. XV.

44. Models with $Q \neq 0$

In the flow (12a) past a plate in a channel, one can consider Q as positive by regarding the upstream velocity (unity, by choice of units) as

[9] The derivation of the Brillouin condition from (14) was suggested in [14], p. 51; see also §45.

the free stream velocity. With this convention, $Q = v^{-2} - 1 > 0$, and $1 + Q = v^{-2}$. The admonition following (12b), to base C_D on the free stream-line velocity $v = v_f$, is thus equivalent to the use of the factor $(1 + Q_w)$ in the preceding section.

However, the construction of Helmholtz flows with $Q \neq 0$ in an infinite stream is much more difficult. Moreover, real cavities are of finite size, and the construction of Helmholtz flows having finite cavities is especially difficult because of the

Brillouin Paradox. Finite cavities satisfying the Brillouin Condition are mathematically impossible.

We sketch the proof (see [16]). Since the pressure in the cavity is a minimum (Brillouin Condition), the free streamlines must curve towards the cavity, which must thus be convex. But such a cavity must have a stagnation point where the two free streamlines meet, at minimum pressure p_s. By the Bernoulli Theorem this implies $|u|^2 = 2(p_s - p)/\rho \leq 0$ everywhere except on the free boundary, which implies $u = 0$ identically.

To avoid the Brillouin Paradox, various modified Helmholtz flows have been constructed, involving artificial alterations of the rear of the cavity. These can be made to yield $Q \neq 0$ without, one may presume, seriously affecting the flow near the obstacle creating the cavity.

Thus, in 1921 Riabouchinsky [37] constructed Helmholtz flows with free streamlines about two symmetrically placed plates (see Fig. 15a), with $Q > 0$. The construction may be sketched as follows (cf. [15], Ch. V, §9).

The "hodograph" (i.e. diagram in the ζ-plane) of one quarter of the flow is clearly a quarter-circle, and the W-domain is a quadrant, by vertical and horizontal symmetry. Hence the conformal map of the hodograph on the W-domain satisfies, much as in §40,

$$(16a) \qquad W^2 = \frac{a(\zeta^2 + \zeta^{-2}) + b}{c(\zeta^2 + \zeta^{-2}) + d} = C \frac{\zeta^4 + 2\lambda\zeta^2 + 1}{\zeta^4 + 2\mu\zeta^2 + 1}.$$

By choice of units we can reduce to the case $C = 1$, and $\zeta = 1$ at $W = \infty$, which implies $\mu = -1$. This gives

$$(16b) \qquad W = \frac{\sqrt{\zeta^4 + 2\lambda\zeta^2 + 1}}{(\zeta^2 - 1)}.$$

from which $z = \int dW/\zeta$ can be obtained as an elliptic integral. If v is the free streamline velocity, then $\zeta = v$ when $W = 0$, whence $\zeta^4 + 2\lambda\zeta^2 + 1 = (\zeta^2 - v^2)(\zeta^2 - v^{-2})$, and $\lambda = -\frac{1}{2}(v^2 + v^{-2})$. By Bernoulli's equation, $Q = v^2 - 1$, whence $\lambda = \frac{1}{2}[(1 + Q) + (1 + Q)^{-1}]$. With these formulas it is easy to obtain C_D as a function of Q.

Another important Helmholtz flow with $Q > 0$ was constructed in 1946 by Efros, and independently by Gilbarg and Rock.[10] This involves a *reentrant* jet (see Fig. 15b) instead of a symmetric cavity. Reentrant jets have been observed experimentally, though they appear to form intermittently and to be unstable.[11] Hence the re-entrant jet model has a special physical interest.

One can also construct "cusped" cavities with $Q < 0$ behind convex bodies—though it was once believed that this was impossible.[12] However,

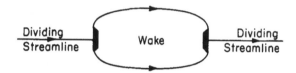

FIG. 15a. Riabouchinsky flow

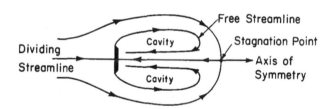

FIG. 15b. Reentrant Jet

these do not at all resemble physical cavities, of which samples will be shown in Plates I-II (see §51). It would seem to be very difficult to maintain overpressure in a stable cavity (or finite wake).

45. Curved barriers

The techniques described above depend basically on the recognition of *special* conformal transformations and on integration formulas for *special* functions. Though they have been skillfully developed and are adequate for handling many problems involving polygonal barriers (see [15], Chs. II, III, and V), they are generally inadequate for the analysis of cavity flows past *curved* barriers.

The development of high-speed computing machines has opened the door to an entirely different approach, based on *general* function-theoretic

[10] See [15], Ch. III, §8 for references and detailed formulas.
[11] Also, at the rear of the cavity behind an axially symmetric obstacle, a pair of hollow core vortices may form ([39], p. 230).
[12] See [15], Ch. V, §§10, 14; also [14], p. 58.

methods. Though this approach has so far been successfully applied only to *plane* flows, and though I shall treat only such applications below, similar techniques should be applicable also to axially symmetric and even arbitrary flows with free boundaries.

We shall introduce this modern approach by considering a general curved barrier held symmetrically in an infinite stream, as in Fig. 16. We shall again let the wetted sector $\widehat{ACB}$ of the barrier be vertical and choose units so that $|\zeta| = 1$ on the free boundary.

Following Levi-Civita [35], we shall map the simply-connected interior of the flow conformally and one-one onto the interior of the semicircle

(17) $$\Gamma: \quad |t| < 1, \quad \operatorname{Im}\{t\} > 0.$$

By the Fundamental Theorem of Conformal Mapping,[13] there is exactly one such mapping $t = f(z)$ of the flow onto Γ, which takes A, B, C into $1, -1$, and i respectively. This $f(z)$ will clearly map the *free* streamlines

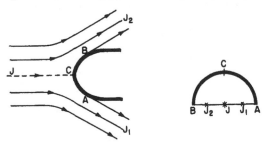

FIG. 16. Flow past curved barrier

onto the real *diameter*, and the wetted portion of the *barrier* onto the semicircle $t = e^{i\sigma}$ $(0 < \sigma < \pi)$. (In this respect it is the opposite of the parametrizations of §§38–40.)

To express the complex potential, it is convenient to map Γ on the upper half-plane by the conformal transformation

(18) $$T = -\frac{t + t^{-1}}{2}, \quad \text{so that} \quad \frac{dT}{dt} = -\frac{1 - t^{-2}}{2}.$$

The complex potential is then evidently given by

(18′) $$W = \frac{MT^2}{2}, \quad \frac{dW}{dT} = MT, \quad M > 0$$

for some positive constant M. For, (18′) maps the flow onto a slit plane, with the dividing point $t = i$ at $W = T = 0$, and the point at infinity $t = 0$ at $W = T = \infty$.

---

[13] L. Bieberbach, *Lehrbuch der Funktionentheorie*, Springer, Leipzig, 1923, vol. *1*, p. 61. For Schwarz' Reflection Principle, see ibid., p. 225.

Now consider the function $(i-t)/(i+t)=(1+it)/(1-it)$. It has modulus one when t is real; its argument in $\pi/2$ on $\widehat{AC}$ and $-\pi/2$ on $\widehat{CB}$, with a jump of $-\pi$ at C. The new function $\Omega(t)=\theta+i\tau$, defined by

$$(19) \qquad \zeta = \left(\frac{1+it}{1-it}\right)e^{-i\Omega(t)}, \qquad \zeta^{-1} = \left(\frac{1-it}{1+it}\right)e^{i\Omega(t)},$$

is also analytic and regular inside Γ. On the free boundary, when t is real, $|1+it|=|1-it|$, and so

$$(19') \qquad\qquad 1 = |\zeta| = \left|\frac{1+it}{1-it}\right|e^{\tau(t)} = \tau(t).$$

Hence $\tau(t)$ vanishes on the diameter of Γ: $\Omega(t)$ is *real when t is real.*

It follows, by Schwarz' Principle of Reflection,[13] that $\Omega(t)$ can be extended to an analytic function, regular inside the unit circle $|t|<1$. We can therefore write, in the symmetric case under consideration (ζ and $i\Omega(t)$ real on the imaginary t-axis of symmetry),

$$(20) \qquad\qquad \Omega(t) = a_1 t + a_3 t^3 + a_5 t^5 + \cdots, \quad \text{all } a_i \text{ real,}$$

where the radius of convergence of (20) is at least unity. On the fixed boundary $|t|=1$, somewhat delicate considerations permit one to prove that $\Omega(t)$ is even *continuous* ([15], Ch. VI, p. 135). Informally speaking, the factor $(1-it)/(1+it)$ "takes out" the simple pole in ζ^{-1} (the zero in ζ) at the stagnation point.

Conversely, given a function (20) with radius of convergence one or more, equations (19) and

$$(21) \qquad z = \int_i^t \zeta^{-1}\left(\frac{dW}{dT}\right)\left(\frac{dT}{dt}\right) dt = \frac{M}{4}\int_i^t \zeta^{-1}(t-t^{-3})\, dt$$

define a symmetrically divided flow past a smooth barrier $\widehat{ACB}$, having a continuously turning tangent. This gives Levi-Civita's classic result:

Theorem 1. The symmetrically divided flows past symmetric barriers in an infinite stream correspond one-one to choices of functions (20), regular for $|t|<1$ and continuous on $|t|=1$, and of constants M. The correspondence is given by equations (19), (20), and (21).

46. Direct problem

Theorem 1 solves the *inverse* problem, of describing the class of *all* plane flows of an infinite stream divided symmetrically by a curved barrier. We now turn to the direct problem: given a particular two-dimensional barrier P held symmetrically in an infinite stream, what is $\Omega(t)$? We shall show that this problem is equivalent to a nonlinear integral equation.

In principle it is a straightforward matter to express all properties of the flow in terms of $\Omega(t)$. Thus, along the fixed boundary P ($t = e^{i\sigma}$ in the t-plane), we have for $\Omega = \theta + i\tau$

(22a) $\quad \theta = a_1 \cos \sigma + a_3 \cos 3\sigma + a_5 \cos 5\sigma + \cdots$

(22b) $\quad \tau = a_1 \sin \sigma + a_3 \sin 3\sigma + a_5 \sin 5\sigma + \cdots$ $\left.\right\}$ on $\quad t = e^{i\sigma}$.

It will be convenient to consider also the derivative

(22c) $\quad \lambda(\sigma) = -d\theta/d\sigma = a_1 \sin \sigma + 3a_3 \sin 3\sigma + a_5 \sin 5\sigma + \cdots$,

assuming, by Hypothesis (E) of §1, that the above Fourier series converge satisfactorily for suitably "smooth" obstacles.

We shall now show that θ differs by $\pi/2$ from the direction ϕ tangent to the obstacle. Since $\arg dz/dT = \arg \zeta^{-1} + \arg dW/dT$ and $\arg dz/dT = \phi$ (except at the stagnation point C), evidently $\arg \zeta^{-1}$ is $\phi - \pi$ on $\widehat{AC}$, and ϕ on $\widehat{CB}$. On the other hand, by (19) and the discussion of $\arg [(1+it)/(1-it)]$ preceding it, $\arg \zeta^{-1}$ is $\theta - \pi/2$ on $\widehat{AC}$ and $\theta + \pi/2$ on $\widehat{CB}$. Combining these two relations, we conclude $\theta = \phi - \pi/2$ throughout $\widehat{ACB}$.

Arc-length ℓ along the barrier can be found by evaluating (21), which gives $d\ell = |\zeta^{-1}| \cdot |dW/dT| \cdot |dT/dt| \cdot d\sigma$ on $t = e^{i\sigma}$. By (19'), and elementary trigonometry in the complex plane,

$$|\zeta^{-1}| = \left|\frac{1-it}{1+it}\right| \cdot |e^{i\theta - \tau}| = \left|\frac{1+\sin \sigma}{\cos \sigma}\right| e^{-\tau}.$$

Similarly, since $dW/dt = (t - t^{-3})$ as in (21),

$$\left|\frac{dW}{dt}\right| = M \cdot |\cos \sigma \sin \sigma|, \quad M = \text{constant}.$$

Combining the above formulas, we get finally

(23) $\qquad d\ell = M\nu(\sigma)e^{-\tau(\sigma)}d\sigma, \quad \nu(\sigma) = |\sin \sigma (1+\sin \sigma)|$.

The *curvature* $\kappa = -d\phi/d\ell = -d\theta/d\ell$ therefore satisfies

(24) $\qquad\qquad\qquad \kappa = \dfrac{\lambda(\sigma)e^{\tau(\sigma)}}{M\nu(\sigma)},$ $\qquad$ (cf. (22c), (23)).

Now let P be any smooth symmetric barrier having *curvature of constant sign* (i.e. no points of inflection), and let $\kappa = K(\theta)$ express the curvature as a function of the angle $\theta = \phi - \pi/2$ through which the tangent has turned after C. Then, transposing (24), we have

(25) $\qquad\qquad \lambda(\sigma) = M\nu(\sigma)e^{-\tau(\sigma)}K(\theta) = M\nu(\sigma)e^{-D\lambda}K(J\lambda)$,

where the linear *operators* D and J define $\tau(\sigma)$ and $\theta(\sigma)$ from $\lambda(\sigma)$, by formulas (22a)–(22c). (Actually, $J\lambda = -\int_{\pi/2} \lambda(\sigma)\,d\sigma$, while $D\lambda$ is the "Dini transform" of $\lambda(\sigma)$ by a suitable singular integral kernel $D(\sigma, \sigma')$; see [15], p. 136.) In conclusion, we have proved

Theorem 2. For a flow given by Theorem 1 to involve a barrier having curvature $\kappa = K(\theta)$ of constant sign, it is necessary and sufficient that (25) hold.

For small M, (25) can be solved by direct iteration of the functional transformation

$$(26) \qquad \lambda_{n+1}(\sigma) = S[\lambda_n(\sigma)] = M\nu(\sigma)e^{-D\lambda_n}K(J\lambda_n).$$

For larger M, "averaged iteration" with respect to a suitable weight factor ϵ converges—i.e. one can iterate

$$(26') \qquad \lambda_{n+1}(\sigma) = (1-\epsilon)\lambda_n(\sigma) + \epsilon S[\lambda_n(\sigma)].$$

Thus, using modern high-speed computing machines, one can effectively solve (25) for given positive M; the details may be found in the literature.[14]

47. Indeterminacy of separation

Because of the parameter M, the correspondence between integral equations (25) and barriers P is not one-one. This raises a basic question: in what sense, if any, is the Helmholtz problem of §36 well-set? This difficult question is still not entirely resolved, even for plane flows having an axis of symmetry.

Thus, as shown by H. Villat [27] in 1911, even the Kirchhoff flow of §39 is not the only possible solution of the Helmholtz problem for a flat plate in an infinite stream. The configuration of Fig. 17 gives rise to a one-parameter family of other, topologically different (Hypothesis (D) of §1) possibilities.[15]

In the case of round obstacles, even if the flow topology is assumed to involve a single infinite cavity behind the obstacle, a more fundamental indeterminacy arises. As first realized by M. Brillouin before any rigorous theorems were available, the *separation point is indeterminate*. This fact is closely related to the indeterminacy of M in (25): in a general way M corresponds to the "wetted length" from the dividing point C to the separation points A and B, and increases with $\overset{\frown}{CA} = \overset{\frown}{CB}$. Thus, the Helmholtz problem is *not* well-set for round obstacles, even if the flow topology is assumed.

---

[14] [15], Ch. IX, §8; G. Birkhoff, H. H. Goldstine, and E. H. Zarantonello, *Rend. Sem. Mat. Torino, 13* (1954), 205–23.

[15] See [15], Ch. V, §3. E. H. Zarantonello, *J. de Math., 33* (1954), 29–80, has shown that there are no other possibilities.

However, if the Brillouin Condition (§43) is assumed, the infinite-cavity problem becomes well-set in at least some cases. Following Leray [34], we define an "accolade" as a barrier P whose curvature $\kappa(\theta)$ is increasing,[16] as in Fig. 18. Leray has proved that any symmetric accolade P has a unique pair of "Brillouin points" A_0, B_0 with the following property: The curvature of the free streamlines at the separation points A, B of any

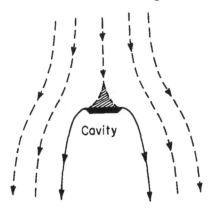

Fig. 17. Villat flow past flat plate

symmetric flow past part of P is $+\infty$, finite, or $-\infty$, according as separation is forward of A_0, B_0, at A_0, B_0, or behind A_0, B_0, respectively. In the case of forward separation, the infinite curvature forces the free streamlines to penetrate the accolade, which is absurd. In the case of separation behind A_0, B_0, the Brillouin Condition is obviously violated. Hence, if we define the *Helmholtz-Brillouin problem* as that of finding

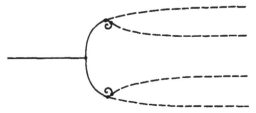

Fig. 18. Helmholtz flow past accolade

Euler flows whose boundary consists of portions of a fixed contour, plus free streamlines satisfying the Brillouin Condition, we have the following conclusion:

For infinite symmetric cavities behind accolades, the Helmholtz-Brillouin problem is well-set, and separation occurs at the Brillouin points

---

[16] By the four-vertex theorem the circle is the only "accolade" which bounds a smooth convex domain.

A_0, B_0. It would be interesting to determine the exact class of symmetric barriers for which the Helmholtz-Brillouin problem is well-set.

In the case of accolades, we have shown above that the Brillouin Condition is equivalent to that of finite curvature of the free streamline. In the literature the latter condition (of "smooth separation"—see [15], Ch. VI, §6) has been repeatedly urged as a "physically reasonable" substitute for the Brillouin Condition. However, in view of (14), especially since the Kirchhoff flow past a flat plate violates the condition of "smooth separation," the Brillouin Condition seems to me preferable.[17]

48. Axially symmetric Helmholtz flows

Definitive mathematical treatments of axially symmetric Helmholtz flows date from 1946, when Levinson[18] gave a rigorous analysis of the asymptotic cavity profiles. Assuming that these satisfied

$$(27) \qquad y = x^s g(x), \quad \text{where} \quad \lim_{x \to \infty} \frac{xg'(x)}{g(x)} = 0,$$

Levinson proved that $s = 1/2$, and that

$$C(\ln x)^{-1/4-\epsilon} < g(x) < C(\ln x)^{-1/4+\epsilon}$$

for some constant C, all $\epsilon > 0$, and sufficiently large x. If (27) is strengthened to $xg'(x)/g(x) = 0(1/\ln x)$, then

$$(28) \qquad y \sim Cx^{1/2}(\ln x)^{-1/4} \quad \text{(i.e.} \left\{ \lim_{x \to \infty} \frac{y(\ln x)^{1/4}}{\sqrt{x}} \right\} = \text{const.,)}$$

and the drag is given by

$$D = \frac{\pi \rho C^4 v^2}{8}.$$

However, Levinson did not prove the existence of such cavities.

The first proof of the existence of *finite* axially symmetric cavities was given in 1952 by Garabedian, Schiffer, and Lewy [17]. Using Riabouchinsky's principle that free streamlines extremalize induced mass, relative to variations leaving the cavity volume constant—and the new result that "symmetrization" decreases induced mass—these authors proved the existence of axially symmetric Helmholtz flows of the "Riabouchinsky" type discussed in §44, for profiles of general shape (and general $Q > 0$). The existence of Helmholtz flows with infinite axially symmetric cavities, though made plausible, was not proved in detail.

---

[17] See also the discussion of [14], §5, where the questions involved were first discussed from the present viewpoint.

[18] N. Levinson, *Annals of Math.*, 47 (1946), 704–30.

The uniqueness of infinite, axially symmetric cavities, for barriers with given separation point, has been proved by Gilbarg and Serrin.[19] The proof is based on comparison methods first introduced by Lavrentiev.

A remarkable fact about the preceding proofs is their use of new basic ideas. This is necessary because the conformal mapping techniques, traditionally used for plane flows, are no longer available.

Curiously also, though existence and uniqueness of plane flows with free boundaries was first proved more than 50 years after the first non-trivial examples of such flows were constructed, no analytical ("exact") axially symmetric Helmholtz flow of interest is yet known[20]—even though existence and uniqueness theorems are available.

Therefore, discussions of specific axially symmetric Helmholtz flows must rely on approximate methods. Of those applied so far, the most ingenious is that of expansion in powers of the dimension number developed by P. Garabedian [30a]. Whereas earlier authors had obtained contraction coefficients of 0.61 for the jet from a circular hole in a flat plate, Garabedian calculates the value as 0.58.

49. Conservation laws

The proofs of the results stated in §48 are extremely sophisticated mathematically. Useful results about axially symmetric Helmholtz flows are often obtained by a much simpler appeal to physical conservation laws, as we shall now see.

Borda tube. Let a container with vertical walls be filled with a liquid of density ρ, into which is inserted a horizontal "Borda tube" of arbitrary cross-section shape and area A (see Fig. 19); let the pressure head at the level of the tube be p. We assume that the flow separates[21] from the tube at its inner end, and that the jet from the tube tends asymptotically to a constant speed v; this must be the constant "free streamline" velocity on the jet boundary. Let A^* be the asymptotic jet cross-section; then A^*/A is by definition the coefficient of contraction. We compute it as follows:

Per unit time, the volume efflux is vA^*; the momentum efflux is v^2A^*; the kinetic energy efflux is $1/2\rho v^3 A^*$. The momentum supplied, however, is pA (pressure excess); the (potential) energy is $p(vA^*)$. Hence $pA =$

---

[19] D. Gilbarg, *J. Rat. Mech. Anal.*, *1* (1952), 309–20; and J. B. Serrin, ibid., *2* (1953), 563–75; see also [15], Ch. IV, §§12–14. For Lavrentiev's work, see *Mat. Sbornik*, *46* (1938), 391–458.

[20] For a discussion of approximate solutions, see G. Birkhoff, Symposium on Naval Hydrodynamics, Aug. 25–29, 1958, [40].

[21] This is the case of "free flow"; see A. H. Gibson, *Hydraulics*, Constable, London, 4th ed., p. 122; it is not always fulfilled.

pv^2A* and $pvA* = 1/2\rho v^3A*$. Dividing v times the first equation by the second, we get the result

$$(29) \qquad\qquad \frac{A*}{A} = \frac{1}{2}.$$

Shaped charges. Another important application of conservation laws is to _lined-cavity charges_, such as were used in the American "bazooka," the British P.I.A.T., and other antitank and demolition weapons of World War II. I shall briefly summarize this application of free streamline theory; further references may be found in [15], p. 16.

The construction and performance of such weapons may be described in principle as follows. An explosive charge, surrounding a hollow "cavity" lined with metal, is detonated from the rear. I shall consider only the

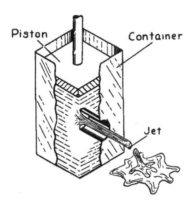

Fig. 19. Efflux from Borda tube

cases of conical and wedge-shaped liners; a sectional view of these is given in Fig. 20a. As the charge explodes, it drives the liner inward and forward, and it is found that the liner so impelled has an enormous penetrating power. How can we explain this power?

The best known explanation proceeds from the following plausible approximate assumptions: Assumption 1: After receiving an initial impulse from the explosive, the liner walls move inwards under their own momentum with constant velocity until they meet on the "axis" (AA' of Fig. 20a). Assumption 2: Under the terrific stresses generated, the liner behaves like a perfect fluid. Assumption 3: Relative to axes moving with the junction J of opposite walls, the flow is stationary. Assumption 4: The surfaces of the liner walls are free boundaries.

These assumptions reduce the analysis to a Helmholtz problem involving "impinging jets" (Fig. 20b). In the plane (wedge) case, the hodograph is a circle; the hodograph of half the flow is a semicircle. The W-domain

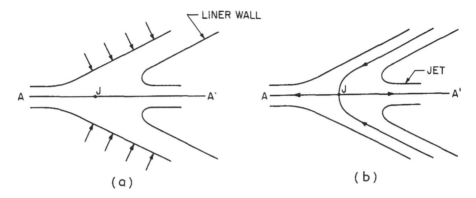

FIG. 20. Jets from shaped charges

is a cut infinite strip, and the method of §37 permits one to calculate the complete flow pattern.[22]

No such technique is available in the more important case of conical liners. However, conservation laws still enable one to predict the jet velocity and mass approximately, as functions of the cone vertex angle and explosive used. Knowing these, the penetration can be estimated from the Bernoulli equation.[23]

50. Cavities as Helmholtz flows

In §43, a rational argument was given for the empirical dictum (Betz-Petersohn[7]) that free streamline theory should be applicable if $\rho'/\rho \ll 1$. This suggests the possibility that cavity flows can be predicted mathematically, as solutions of the Helmholtz-Brillouin boundary-value problem. Evidence for and against this conjecture will now be reviewed, only favorable evidence being presented in this section.

First, as is the case with ideal cavity flows, real cavity profiles are relatively smooth, stationary,[24] and ten or more body diameters long. Thus, they resemble the theoretical model infinitely better than do real wakes (see §53). A minor exception is provided by obstacles held in water tunnels, with $Q > 0.3$.

Second, the cavity profile is nearly always convex, and flow separation occurs in front of the cross-section of maximum diameter. This again

---

[22] [15], p. 36; [8], p. 283.

[23] G. Birkhoff, D. P. MacDougall, E. Pugh, and Sir Geoffrey Taylor, *J. Appl. Phys.*, *19* (1948), 563–82.

[24] Unless the obstacle is extremely smooth, small "flutings" parallel to the flow are apt to occur, as in Plate I or [38], pp. 116, 129. When $Q > 0.3$, the cavity may be alternately filled and emptied of water; see [40], p. 10.

agrees generally with solutions of the Helmholtz-Brillouin problem, and is in notable contrast with the case of wakes.

Third, the approximate experimental formula[25]

$$(30) \qquad C_D = 0.55 + 0.4Q = 0.55(1 + 0.73Q),$$

for the cavity C_D of a cylinder broadside, agrees quite well with Brodetsky's theoretical value $C_D(0) = 0.55$ for the Helmholtz-Brillouin problem (calculable by the method of §46), when corrected by the factor $(1+Q)$ of §43.

Some of the most interesting evidence concerns the cavities behind missiles shot into water. Plate I is a sample piece of evidence[26]; it shows the cavity formed behind a sphere entering the water at about 150 ft/sec. Two exposures, separated by about 0.005 secs., are made on a single plate. The white dots speckled over the picture are small bubbles, each bubble being photographed twice. It is not hard to pair the bubbles, and the vector from the first to the second bubble of each pair is roughly proportional to the vector velocity of the water in the vicinity. In this way the cavity profile and the velocity field of the flow can be clearly visualized.

Now, photographs such as these are not exactly comparable with theory, for various reasons. Thus they represent a decelerating body, and not stationary flow; the water surface presents a second free boundary, which complicates the mathematical descriptions; moreover the influence of air is not negligible (cf. §53).

Qualitatively, the applicability of the theory of Helmholtz flows is supported by the fact that arbitrarily long (100 diameters or more) cavities occur behind missiles moving sufficiently fast. This fact has a considerable practical importance, in that much of the wounding effect of high-speed bullets and bomb fragments is due to their ability to create temporary holes much bigger than themselves.[27] For us, its significance is that it provides physical approximations to the infinite cavities defined mathematically as solutions of the Helmholtz-Brillouin problem.

Extended Helmholtz Problem. If one assumes Eq. (14) and the properties of incompressible nonviscous fluids, one can apply the concepts of Helmholtz also to *accelerated flow*, involving *gravity* forces. To do this, one assumes that cavitation occurs spontaneously whenever $p < p_v$. The

---

[25] H. Kempf and E. Foerster, eds., *Hydrodynamische Probleme des Schiffsantriebs*, Springer, 1932, 227–342.

[26] See G. Birkhoff and T. E. Caywood, *J. Appl. Phys.*, *20* (1949), 646–59, for the experimental technique used and other photographs.

[27] See [3], Sec. 74, and refs. given there; also E. N. Harvey et al., *The Military Surgeon*, *98* (1946), 509–28.

resulting boundary-value problem may be called the *extended* Helmholtz problem.[27a]

The idea that real cavitation can be mathematically predicted by examining solutions of the extended Helmholtz problem is supported by the qualitative observation that *vapor cavities start at solid surfaces*. This empirical principle can be deduced within the framework of the extended Helmholtz problem as follows.[28] Taking the Laplacian of Bernoulli's equation (Ch. I, (5)), we have

$$(31) \qquad \nabla^2 p = -\rho_0 \nabla^2 \left\{ \tfrac{1}{2} \nabla U \nabla U + \frac{\partial U}{\partial t} + G \right\}.$$

In (31), $\nabla^2 G = 0$, since G is the Newtonian gravitational potential; $\nabla^2(\partial U/\partial t) = \partial(\nabla^2 U)/\partial t = 0$ by Ch. I, (6); and, setting $u_k = \partial U/\partial x_k$ so that $\nabla U \nabla U = \sum u_k^2$,

$$\nabla^2 (\sum u_k^2) = \sum u_k \nabla^2 u_k + 2 \sum (\nabla u_k \cdot \nabla u_k) \geq 0.$$

Hence $\nabla^2 p \leq 0$, with equality holding only if u is constant; that is, p is *superharmonic*. It is well known, however, that a superharmonic function must assume its minimum values on the boundary; hence p must first fall below p_v there.

51. Bubbles

Often used interchangeably for "cavity," the word "bubble" tends to connote smallness and mobility. In treating small bubbles, it is typically necessary to consider the effects of gravity and surface tension—as we have already seen in §32. We shall now present more results about bubbles which further illustrate this principle—and thus reveal clearly ways in which Helmholtz flows give only an approximate picture of real cavities.

First, we recall ([11], vol. 1, p. 62) the jump $2\gamma/r$ in pressure from the outside of a spherical bubble of radius r to the inside, due to the surface tension γ. This superficial consideration indicates the possibility that a liquid, from which all bubbles of radius $r > R$ have been removed, can support a tension of $(2\gamma/r) - p_v$ without cavitation!

Though limitations of space prevent our exploring this fascinating question in detail, we do recall that tensions of tens of atmospheres[29] have been sustained by liquids, under controlled laboratory conditions

---

[27a] The theory of gravity waves treats a closely related problem in which $p = p_a$ on a free surface. Ordinarily, $p > p_a$ under the surface, but this condition is not assumed.

[28] G. Kirchhoff, *Vorlesungen über Mechanik*, 1876, p. 186; see also G. Bouligand, *J. de Math.*, 6 (1927), p. 427.

[29] On the other hand, the figure of 250 atmospheres often quoted seems incorrect; see [15], Ch. XV, §3.

involving degassing, in spite of Eq. (14). Similarly, de-aerated water can be superheated without forming steam. For these reasons, laboratory measurements of cavitation in liquids now include the routine measurement of their air content. It is only because most "water" is *not* nearly homogeneous (cf. §1), but contains many "bubble nuclei" in suspension, that (14) is approximately valid.

A second topic, of mathematical interest, concerns the rise of large bubbles in vertical tubes, under gravity. Without going into the difficult questions of physical realization and stability, and neglecting surface tension, we shall consider the ideal case of a two-dimensional "rising plane bubble," sketched in Fig. 21a. The point of greatest interest is the close analogy with the mathematical methods introduced in §§45–46.

To show this, we again map the flow onto the unit semicircle Γ in the t-plane (Fig. 21b), but with the fixed boundary mapped onto the diameter and the free boundary onto the semicircumference, as in §37. Let d be the tube diameter and u_0 the rate of bubble rise (fixing axes in the bubble vertex, u_0 becomes the rate of fall at ∞_0, the upstream point at infinity). Then, almost by inspection, the complex potential

$$(32) \qquad W = A \ln\left(\frac{t}{1-t^2}\right), \quad A = \frac{u_0 d}{\pi},$$

puts a source of the right intensity at $t=0$, equal sinks at $t=\pm 1$ (∞_1 and ∞_2), and makes the boundaries of Γ into streamlines.

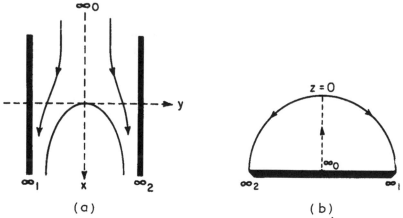

FIG. 21. Rising plane bubble in channel

As to the conjugate velocity $\zeta(t)$, we take out its zeros and infinities in Γ by the following analog of Levi-Civita's substitution (19):

$$(33) \qquad \zeta = (1+t^2)[-\ln C(1-t^2)]^{1/2}e^{\Omega(t,C)}, \quad 0 < C < 0.5.$$

As before, Schwarz's Principle of Reflection shows that $\Omega(t, C)$ is regular in the unit circle $|t| < 1$, and its Taylor series

$$(33') \qquad \Omega(t, C) = a_0(C) + a_2(C)t^2 + a_4(C)t^4 + \cdots$$

is Abel summable on $t = 1$.

It remains to satisfy the condition $|\zeta|^2 = 2gy$ on the interface, i.e., the Bernoulli equation for a free boundary in steady incompressible non-viscous flow. This condition is equivalent to a nonlinear integral equation in the unknown function

$$\lambda(\sigma) = -\mathrm{Im}\,\{\Omega(t, C)\} = -2a_2 \sin 2\sigma - 4a_4 \sin 4\sigma - \cdots,$$

defined by the coefficients of (33').

This integral equation is analogous to (25) but more complicated. Its approximate numerical solution proved very difficult. The computations led to the conclusion[30] $u_0/\sqrt{gd} = 0.23 \pm 0.01$, which agrees reasonably well with the limited experimental data available.

52. Taylor instability

When $\rho' > \rho$, the gravity term in formula (13) is clearly *destabilizing*. The instability involved is simply that of a pail of water held upside down!

Since the effect of translating a domain with translational acceleration $\boldsymbol{a}$ is equivalent[31] to imposing a gravity field $\boldsymbol{g} = -\boldsymbol{a}$, the preceding result can also be interpreted in accelerated flow, as follows. A plane interface separating two fluids of densities ρ, ρ' is unstable when accelerated from the lighter towards the heavier fluid. This type of instability is called *Taylor instability*.[32]

The two-dimensional instability of perturbations in an initially plane interface is adequately covered by (13), so long as their amplitude remains *infinitesimal*. For initially sinusoidal perturbations the most conspicuous feature of nonlinear Taylor instability is the development of round-ended, upstanding columns separated by falling jets. Curiously, these columns conform approximately to the analysis of "rising plane bubbles" sketched in §51.

Taylor instability affects pulsating spherical bubbles very significantly. Such bubbles play a central role both in cavitation erosion (§42) and in underwater explosions. By assuming spherical symmetry (Hypothesis (C)

---

[30] G. Birkhoff and D. Carter, *J. Rat. Mech. Anal.*, 6 (1957), 769–80; see also P. Garabedian, *Proc. Roy. Soc.*, *A241* (1957), 423–31.

[31] J. L. Synge and B. A. Griffith, *Principles of Mechanics*, 2d ed., McGraw-Hill, 1949, §5.3.

[32] Because its physical significance was first made clear by Sir Geoffrey Taylor *Prac. Roy. Soc.*, *A201* (1950), 192–6. For further discussion and references, see [15], Ch. XI, §§12–13.

again!), Rayleigh[33] obtained simple differential equations for the radius $b(t)$ as a function of time, applicable to both types of bubbles. However, if perturbations of the spherical boundary are expanded in Legendre functions $p_h(\cos \phi)$, the perturbation amplitudes $b_h(t)$ may be shown to satisfy

$$(34) \qquad\qquad b\ddot{b}_h + 3\dot{b}\dot{b}_h - (h-1)\ddot{b}b_h = 0.$$

(This differs from the plane case of (13) through the term $3\dot{b}\dot{b}_h$.) After becoming overexpanded by the momentum of the water driven outwards, underwater explosion bubbles contract again to roughly their initial radius. Near the minimum radius there is a strong deceleration of the inward flow—i.e. an acceleration in the direction of the denser fluid. This obviously makes the spherical interface Taylor-unstable, a fact which greatly weakens successive pulsations of the bubble.

The case of a vapor-filled bubble contracting to a "point," as in idealized cavitation erosion, is more subtle, because there is always acceleration from the denser to the lighter fluid. Nevertheless, there is instability due to negative damping,[34] essentially since $\ddot{b} < 0$.

The preceding discussion not only neglects many physical variables of potential importance (e.g. surface tension), it is also restricted to *infinitesimal* perturbations. Though some progress has been made in analyzing the growth of asphericity of finite amplitude, the nonlinear theory is not yet well understood.

53. Scale effects in water entry

Most of the experimental evidence reviewed above supports the idea that mathematical solutions of the extended Helmholtz problem are approximately applicable to real cavity flows. The exceptions mentioned so far have been associated with the peculiarities of small bubbles.[35] Moreover, the discussion of §43 gives a rational basis to the conjecture that free boundary theory is applicable when ρ'/ρ is small.

If we wish to square this conjecture with the experimental facts, we are forced to say that 0.0013 is not "small." Specifically, there are two hydrodynamical phenomena which occur with entry into water under atmospheric conditions, and not if the air is evacuated. Hence, no mathematical theory which neglects $\rho'/\rho \leq 0.0013$ can correctly predict them.

The more important is the phenomenon of *surface seal*. If a small sphere is dropped into still water at 10–20 ft/sec, the cavity first closes, as in Fig. 22a, in a *deep seal*, whereas at entry speeds of 40 ft/sec or more,

---

[33] [12], vol. vi, p. 504; [7], Sec. 91a; [15], Ch. XI, §§1–3.
[34] G. Birkhoff, *Quar. Appl. Math.*, *13* (1956), 451–3.
[35] See §§32, 51; and [40]. Small bubbles are a subject in themselves!

the cavity first closes at the surface, as sketched in Fig. 22b. A photograph of the velocity-field associated with surface seal is displayed as Plate II. The phenomenon of surface seal was first observed about 1900 by Worthington [35]; later, Mallock[36] noted that the sound emitted by the collapsing cavity was a "plop" with deep seal, and a "plunk" with surface seal.

In 1944 R. M. Davies,[37] following a suggestion of Sir Geoffrey Taylor, showed that surface seal no longer occurs if the air pressure p is sufficiently reduced, even at higher speeds. It was at first conjectured that the occurrence or non-occurrence of surface seal depended on the cavitation number $Q = 2p/\rho v^2$, since Q is basic for many cavitation phenomena. But

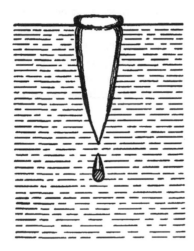 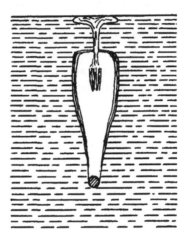

FIG. 22a. Deep seal   FIG. 22b. Surface seal

in 1945, following my suggestion, Gilbarg and Anderson [31] used heavy gases of the Freon series, thereby varying density and pressure independently. They showed that, for a given sphere diameter d and *vertical* entry speed v (Froude number $F = v^2/gd$), the manner of seal was largely determined by ρ'/ρ. For example, with 1-inch diameter spheres entering at 50–150 ft/sec (i.e. $F = 10^3$–10^4), surface seal occurs if $\rho'/\rho > 0.001$; deep seal occurs if $\rho'/\rho < 0.0001$.

This sudden change of regime near a very small value of the parameter ρ'/ρ recalls the sudden changes in ordinary wakes which occur near $1/R = \nu/vd = 0.02$ (alternating vortices) and $1/R = 0.000005$, and in pipes

[36] W. Mallock, *Proc. Roy. Soc.*, *A95* (1918), 138–43. See also E. N. Harvey and J. H. McMillen, *J. Appl. Phys.*, *17* (1946), 541–55.

[37] Unpublished mimeographed report.

near $1/R = 0.0005$ (see Chapter II). Thus it gives us another "paradox of approximation," and illustrates again the principle that *the nature of the solutions of partial differential equations may change abruptly near extremely small values of the parameters.*

It will be shown in §78, though by a somewhat arbitrary physical analysis, that the change of regime is actually associated closely with the dimensionless parameter $N = \sqrt{F}\rho'/\rho$, and occurs when N is about $1/80$.

Another phenomenon, incompatible with naive acceptance of the dictum of Betz and Petersohn, is a tendency to downward refraction in oblique entry. Though the exact facts are not clear, it has been shown experimentally by Prof. L. B. Slichter that a smooth 2-inch diameter dural sphere, entering the water at 50 ft/sec, and at an angle of 20° with the horizontal, may be refracted downwards 5° or more as it enters. (At much higher speeds an upwards deflection and tendency to ricochet is found.[38]) Though a complete theory is not available, Slichter made a careful (unfortunately unpublished) experimental analysis showing that this downward refraction is associated with *air viscosity*—a variable whose effect would intuitively seem negligible (cf. Hypothesis (A) of §1).

54. Real wakes

Although Euler's equations are ordinarily approximately applicable to steady flow conditions when $\rho'/\rho \ll 1$, it is *not* sufficient for ν to be small. This is shown dramatically by photographic studies of real wakes. In particular, the primary variable determining the behavior of real wakes is the dimensionless Reynolds number $R = vd/\nu$, defined in §21. The question is, then, as to the nature of real wakes when $R \gg 1$.

For $R \ll 1$, real wakes have approximately fore-and-aft symmetry, and agree with Stokes' "creeping flow" approximation (§30), when solutions of this boundary-value problem are available. In the range $5 < R < 30$, approximately,[39] the flow past a circular cylinder or other non-streamlined ("bluff") obstacle "separates," and the separating streamlines enclose a finite convex "wake," which resembles qualitatively the finite cavities described earlier in this chapter. Indeed, behind spheres and discs such wakes have been observed up to $R = 200$.

At larger R, but most typically in the range $40 < R < 1000$, real wakes are ordinarily *periodic*, often giving rise to an audible musical note. The frequency N of vibration is related to the flow speed v and diameter d, in the case of a circular cylinder, by the approximate empirical equation

$$(35) \qquad\qquad N = \frac{v}{6d}.$$

---

[38] C. Ramsauer, "Uber den Ricochetschuss", Kiel dissertation, 1903 ; [3], p. 453.
[39] See [15], Chs. XII–XIV for a more detailed discussion of the facts.

Since the flows inducing such periodic wakes are steady, this represents a symmetry paradox (§26).

In the range $10^3 < R < 10^5$, real wakes behind bluff bodies are dominantly *turbulent*, but the boundary layer does not usually become turbulent until after separation in the case of reasonably smooth surfaces. For $R > 3 \times 10^5$, however, the boundary layer normally becomes turbulent *before* separating. As already explained in §28, this gives rise to a greatly narrowed (still turbulent) wake.

The essential dependence of all these qualitative phenomena on the numerical magnitude of R makes it evident that no really fundamental theory of real wakes can ignore viscosity. Nevertheless, various ingenious models for wakes have been constructed using Euler's equations.

Thus, von Mises[40] has suggested that the solution of the Helmholtz-Brillouin problem should be taken as a reasonable approximation to the real flow with laminar boundary layer past a cylinder—whereas, as the wake contracts at higher Reynolds numbers due to boundary-layer turbulence, the extreme "wake of zero drag" gives a good approximation.

Other "free streamline" models have been proposed,[41] in which there is a partial pressure recovery in the wake at large distances, which can be fitted to empirical data.

55. Vortex models for wakes

Still other wake models introduce *ad hoc* vorticity distributions to fit observed flow features by simple mathematical equations. For large R, the neglect of viscosity is again at least superficially plausible.

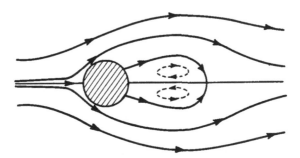

FIG. 23. Föppl flow behind cylinder

One old model involves two symmetrically disposed point-vortices behind a circular cylinder, as in Fig. 23 (cf. Fig. 13). This model, due to Foppl, has received considerable attention because of its extreme mathe-

[40] *Theory of Flight*, McGraw-Hill, New York, 1945, p. 101.
[41] A. Roshko, *J. Aer. Sci.*, *22* (1955), 124–32; W. A. Cornell in [40].

matical simplicity, its interesting stability theory, and its relation to vortex street models (§56). Provided the vortices lie on the curve $2ry = r^2 - a^2$ ([7], p. 223), a being the cylinder radius, the vortex configuration is in (unstable) equilibrium. Moreover, the model can be made to give streamline configurations agreeing well with those observed experimentally for $5 < R < 30$ (say).

However, near a vortex center in steady plane flow, many velocity fields are compatible with the streamline configuration of concentric circles, and so not too much significance should be attached to this agreement. Theoretically, in steady flow the vorticity should diffuse outward from any center, becoming asymptotically *constant* inside any closed streamline in laminar flow at high Reynolds number.

This much more plausible model has recently been proposed by Batchelor,[42] but the computational difficulties of working out specific predictions have not yet been surmounted. Moreover, owing to Helmholtz instability, real wakes at large R involve highly unsteady flow, and so the model is unrealistic. For the range $5 < R < 30$, Oseen's viscous flow model (§31) seems more plausible.[43]

56. Vortex streets

The most intriguing vortex model for wakes is the "vortex street," consisting of two parallel rows of equally spaced point-vortices, these periodic arrays being "staggered" so that vortices in each row are midway between those in the other. This model was proposed by von Karman[44] as a representation of the periodic wakes behind cylinders observed in the range $30 < R < 300$, especially. Its complex potential $W = U + iV$ is

$$(36) \qquad W = \frac{i\kappa}{2\pi}\left\{\log \sin \frac{\pi z}{a} - \log \sin \frac{\pi}{a}\left(z - \frac{a}{2} - ih\right)\right\}.$$

Thus three parameters are involved: the vortex strength κ, the longitudinal spacing a, and the transverse spacing h. The determination of these parameters in any given case is clearly basic; any choice of κ, a, and h gives an array which is in equilibrium.

Von Karman showed that in a nonviscous fluid such an array has first-order instability (i.e. the deviations from equilibrium grow exponentially), unless $h/a = 0.281$ (approximately). He also showed that the analogous array, in which the vortices in the two rows were parallel (the $a/2$ being omitted in (36)), was always unstable.

[42] *J. Fluid Mech.*, *1* (1956), 177–90 and 388–98. See also W. W. Wood, ibid., *2* (1957), 77–87.

[43] References are given in [15], p. 263, ftnt. 13. An excellent historical review of vortex systems in wakes has been made by L. Rosenhead, *Adv. Appl. Mech.*, *3* (1953), 185–98.

[44] *Gott. Nachr., Math.-Phys. Kl.*, (1912), 547–56.

Moreover, by considering the rate K at which vorticity was shed on each side in the boundary layer, Heisenberg and Prandtl[45] have shown that

$$(37) \qquad \frac{\kappa v}{a} \simeq K \simeq \frac{(1+Q)v^2}{4},$$

whence $\kappa \simeq (1+Q)av^2/4$. Though the argument is very approximate, the conclusion is reliable up to a factor of two. Finally, one can easily guess that h will not differ too much from the cylinder diameter d; this guess will be given a more logical basis in §57. By combining the above arguments, an approximate *a priori* model can be constructed for periodic wakes.

It is obvious that this vortex street model does not arise from the solution of a mathematical boundary-value problem: von Karman's ingenious idea is not part of "rational hydrodynamics" in the sense of §1. Thus the obstacle does not appear as a geometric entity in the theory.

It has been suggested that vortex streets arise naturally from the rolling up of vortex sheets, thus representing asymptotic solutions of an initial-value problem. However, the approximation of concentrated *point-vortices* is unrealistic, both theoretically and experimentally,[46] even though it is sometimes said that vortex sheets roll up, "the vorticity becoming more and more concentrated into the rolled up portions."

These observations are intended to stress how far modern fluid dynamics has evolved from the simple and dogmatic idea of Lagrange. All the steady vortex flows of §55 and all solutions of the Helmholtz problem satisfy Euler's equations for an incompressible nonviscous fluid—which shows how far from being "well-set" is the steady flow problem for these equations. In reality, the whole concept of "steady flow" is erroneous physically for fluids of low viscosity!

57. Wake momentum

However, the fact that Lagrange's ideas were erroneous does not mean that the rational approach to hydrodynamics should be abandoned. As we have seen in Ch. II, there is good reason to believe that the Navier-Stokes equations for an incompressible fluid are trustworthy. Our discussion of wakes will be concluded by a brief survey of what has been achieved to date using these equations. As in the case of cavity motion (§49), much can be learned from interpretation of conservation laws.

[45] W. Heisenberg, *Phys. Zeits.*, 23 (1922), 363–66, and comments on p. 366 by Prandtl. See also [4], pp. 555, 564, and [11], vol. 2, p. 132.

[46] For a theoretical critique see G. Birkhoff and J. Fisher, *Rend. Soc. Mat. Palermo* 8 (1959), 77–90; for experimental data see [4], Ch. XIII. The quotation is from [4], p. 62.

Perhaps the most useful interpretation concerns *wake momentum*. Experimentally, one observes a forward-moving "wake" behind any solid moving through a fluid. Thus, in two dimensions one can define the "wake momentum per-unit-length" at a distance x behind an obstacle by the formula ([15], p. 266)

$$(38) \qquad\qquad M(x) = \rho \int u(x, y)\, dy,$$

where the x-axis is made parallel to the direction of motion of the solid, and u denotes the forward velocity[47] of the fluid.

If we admit the empirical fact that the vorticity is negligible outside the wake, so that the approximations of classical hydrodynamics are applicable there, then we are led to the conjecture that $M(x)$ should become virtually independent of x, a short distance behind any obstacle.[48] This hypothesis is also confirmed experimentally.

Further, it is natural to guess that the wake momentum is created by the thrust of the object on the fluid (Newton's Second Law), and that this is equal and opposite in magnitude to the drag D exerted by the fluid on the object (Newton's Third Law). Specifically, D should equal the wake momentum created per-unit-*time*, and this should equal $u_\infty M$, where M is the asymptotic wake per-unit-*length*.

These intuitive guesses can be formulated mathematically and deduced from reasonable hypotheses regarding the fluid flow.[49] Even more interesting is the fact that a refinement of the formulas gives the best means for *measuring* the real drag of an airfoil in flight, from Pitot pressure traverses made a fraction of a chord-length behind the airfoil.[50]

For us, a still more interesting application of wake momentum conservation is to the "vortex street" model of §56. In a very long vortex trail with bounded velocity, the *average* longitudinal spacing $\bar{a}$ of vortices cannot change with time. On the other hand, the wake momentum per-unit-length is easily calculated as $\bar{h}\kappa/\bar{a}$, where $\bar{h}$ is the average transverse spacing. It follows theoretically that in a nonviscous fluid, when κ is constant in time, $\bar{h}$ (and hence the ratio of mean transverse to mean longitudinal spacing) must be *constant* in time: there is no tendency to a unique, "stable" spacing ratio. In a viscous fluid the concentrations of

---

[47] In periodic or turbulent wakes, the time average of this velocity.

[48] To give such hypotheses mathematical respectability, one can call them "Tauberian."

[49] [15], Ch. XII, §9. An earlier careful discussion is by G. I. Taylor, *Phil. Trans.*, A225 (1925), 238–45; see also S. Goldstein, *Proc. Roy. Soc.*, A142 (1933), 563–73.

[50] [4], §115. The technique was developed originally by A. Betz, *Zeits. Flugt. Motorluftschiffahrt*, 16 (1925) 42–44; see also A. Fage and B. M. Jones, *Proc. Roy. Soc.*, A111 (1926), 592–603.

[51] Actually, even the Kármán spacing ratio $h/a = 0.281\cdots$ is unstable, though to lower order; see [29].

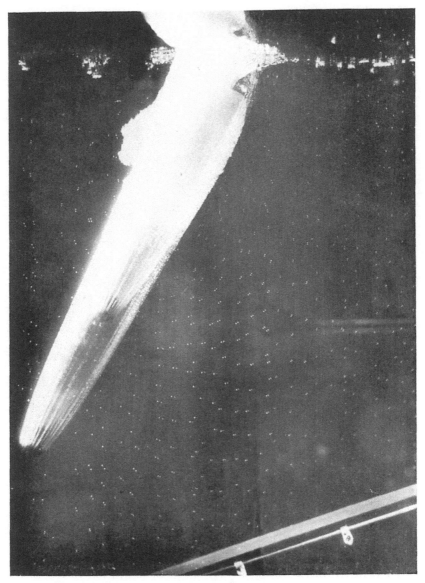

PLATE I. Velocity field and cavity behind high-speed sphere in water.

PLATE II. Velocity field for surface seal at shallow entry angle.

vorticity $\pm\kappa$ of opposite sign diffuse and annihilate each other; hence $\bar{h}$ can be expected to increase; this is observed experimentally.

Most interesting of all, scientifically, is the light which the above result sheds on the asymptotic structure of real wakes. However, to deduce this, one must introduce a further Similarity Hypothesis. Similarity and related ideas will be our main concern in Chs. IV–V. For the applications to wakes, see [15], Chs. XII and XIV.

IV. *Modeling and Dimensional Analysis*

58. On models

The use of *models* to study fluid mechanics has an appeal for everyone endowed with natural curiosity. What active boy has not played with ship and airplane models or crude models of dam and drainage systems? Even in the most advanced technical engineering, such models play a fundamental and indispensable role.

And yet in few departments of the physical sciences is there a wider gap between academic doctrine and engineering practice than in the use of models to study hydrodynamic phenomena. Academic scientists tend to gloss over uncomfortable facts which do not fit nicely into a simple logical theory. Whereas engineers, constantly faced with reality in the field and in the laboratory, are usually too engrossed with technical special problems to enter into the arena of academic controversy. It is easier to pay lip service to current theories, relying on experience and judgment for the solution of design problems.

My aim is to narrow this gap by a critical examination of the whole field from both sides. I shall begin with *theories* of modeling, emphasizing the connection with the *group* concept.

First, in §§59–65, a critical account will be given of *dimensional analysis*. This is usually invoked in justifying model experiments; it has the advantage of requiring no mathematical background beyond high-school algebra, but the disadvantage of needing additional postulates, whose physical validity must be tested independently. In §§60–61, I give these postulates a *group-theoretic* formulation, in terms of the "dimensional group" of all changes of fundamental units.

In §§66–74, I then show how these postulates can be deduced from the mathematical formulations of fluid mechanics discussed in Chapters I–II, by inspecting their equations for invariance under given groups. Borrowing a suggestive phrase coined by Ruark [55], I have called this *inspectional analysis*; it is really an old idea, proposed by Mrs. Ehrenfest [49]. But it has never been technically exploited before; it is shown to be far more reliable (§72) and more general (§74) than dimensional analysis.

Finally, in §§75–8, the deductions from inspectional (and dimensional) analysis are correlated with the *practice* of modeling. As in Chapters I–II, it is shown that reality is much more complex than theory. Engineering practice must take into account many factors which are ignored in mathematical discussions, and it must use many procedures which are not scientifically justified.

59. Dimensional analysis

Dimensional analysis arose from an attempt to extend to physics the Greek concepts of geometrical similarity, ratio, and proportion.[1] It was first applied by Galileo to predict the strength of beams of given material as a function of their linear dimensions. He assumed it as intuitively obvious that, for any given material, failure of a beam occurred when the force per-unit-area (stress) exceeded a certain maximum. He concluded that the safe load per-unit-volume was inversely proportional to the length, and he anticipated various other useful results.

Other applications were given by Mariotte and Newton,[2] but it was Fourier[1] who first stated that there are certain "fundamental units," in terms of which every physical quantity has certain "dimensions," to be written as exponents. To be sure, this idea is almost obvious if one thinks systematically about the "conversion factors" needed to convert physical quantities from one system of units into another.

By simple changes of units, Fourier found it easy to treat the cooling of small spheres and the cooling of the earth by the same analytical formulas. Since we are interested in the relation of logic to fact, it is relevant that Fourier's deductions were badly off because they ignored convection and radioactive heating in the earth's core! Nevertheless, his procedure for eliminating parameters by changes of units is now classic and has been used with notable (if not uniform!) success in many parts of physics. In the hands of Stokes, Savart, Froude, Reynolds, Vaschy, and many others, it has suggested principles of great importance.

Above all, Rayleigh focused attention on the fruitfulness of the procedure as a general method. This led, during the decades 1900–1920, to an intensive discussion by leading physicists of the assumptions tacitly involved, and to a realization of the limitations of dimensional analysis.

Though still frequently ignored in the engineering and popular literature, these assumptions and limitations are clearly stated in the classic monograph of Bridgman [43], to which the interested reader is also referred for many other aspects of the subject. Other authoritative discussions are by Sedov [56] and Langhaar [51].

---

[1] Thus Fourier ([50], Ch. II, Sec. IX) mentioned the Greek awareness of the dimensions of area and volume. Rayleigh always referred to "similitude" and "dynamic similarity." For Galileo, see his *Two New Sciences* (1638), Second Day.

[2] Mariotte, *Traité de la percussion des corps* (1679) and *Traité du mouvement des eaux* (1686); Newton, *Principia Mathematica* (1686), Vol. 2, Sec. 7. See also J. Bertrand, *J. de l'Ecole Polyt.*, *19* (1848), 189–97.

60. The group of dimensional analysis

One of my main purposes will be to found dimensional analysis on postulates referring explicitly to the "dimensional group" of positive scalar transformations of units, mentioned in §58. Though the postulates will be stated abstractly, they will be interpreted through typical examples from fluid mechanics, of which the following is perhaps the simplest.

Example 1. Let us assume (or remember!) that the speed v of a wave in deep water is determined by its length λ and the acceleration g of gravity, so that $v = f(\lambda, g)$. Let us also assume that the relation is the same in all "fundamental units" of length and time. Having made these assumptions, we can argue mathematically as follows.

Let a wave of length λ in a gravitational field of intensity g move with speed v, in a given set of fundamental units. Relative to any new unit of length equal to α old units, and new unit of time equal to τ old units, the wave-length will be $\lambda' = \lambda/\alpha$ and its speed will be $v' = v\tau/\alpha$, while the acceleration of gravity will be $g' = g\tau^2/\alpha$. Choosing $\alpha = \lambda$ and $\tau = \sqrt{\lambda/g}$, we will have $\lambda' = g' = 1$, and so

$$\frac{v}{\sqrt{g\lambda}} = v' = f(\lambda', g') = f(1, 1) = C,$$

where C is the velocity of a wave one unit long in a gravitational field, in the new system of fundamental units. Hence $v = C\sqrt{g\lambda}$ in the new system of fundamental units. But by hypothesis, the function f is the same in all fundamental units; hence $v = C\sqrt{g\lambda}$ also in the given units, where C is a universal constant. (For waves in deep water, $C = 1/\sqrt{2\pi}$.)

The preceding argument can be paraphrased abstractly in terms of the usual concepts of "fundamental" and "derived" dimensionally homogeneous quantities. These concepts can be characterized by the following two postulates, which are little more than definitions. (In Example 1, λ is a "fundamental" quantity, while v and g are "derived" quantities.)

I. There are certain "fundamental quantities" q_i $[i = 1, \ldots, n]$ such that, for any positive real numbers α_i, we can "change units" independently according to the formula

$$(1) \qquad T_\alpha(q_i) = \alpha_i q_i \quad [i = 1, \cdots, n; \alpha_i > 0].$$

(In mechanics, $n = 3$ and the q_i are length, time and mass; in Example 1, $n = 2$ as mass is not involved.)

II. There are "derived quantities" Q_j (such as density, velocity, viscosity, etc.) which are *dimensionally homogeneous*, in the sense that under (1) each Q_j is multiplied by a suitable "conversion factor" given by

$$(2) \qquad T_\alpha(Q_j) = \alpha_1{}^{b_{j1}} \cdots \alpha_n{}^{b_{jn}} Q_j.$$

The exponents b_{jk} are called the "dimensions" of Q_j in the given fundamental units; if they are all zero, then Q_j is called "dimensionless." Clearly, any power-product of dimensionally homogeneous quantities is dimensionally homogeneous. Clearly also, (2) contains (1) as the special case

$$b_{ik} = \delta_{ik} = \begin{cases} 1 & \text{if} \quad k = i \\ 0 & \text{if} \quad k \neq i \end{cases}$$

Hence, mathematically, Assumption I is redundant.

The transformations T_α correspond one-one to vectors $\boldsymbol{\alpha} = (\alpha_1, \ldots, \alpha_n)$ with positive components ("positive" n-vectors). Moreover, if we define

$$(3) \qquad \boldsymbol{\alpha\beta} = (\alpha_1\beta_1, \cdots, \alpha_n\beta_n), \qquad \boldsymbol{\alpha}^{-1} = (\alpha_1^{-1}, \cdots, \alpha_n^{-1}),$$

then evidently

$$(3a) \qquad T_\alpha(T_\beta(Q_j)) = T_\beta(T_\alpha(Q_j)) = T_{\alpha\beta}(Q_j),$$

$$(3b) \qquad T_{\alpha^{-1}}(T_\alpha(Q_j)) = Q_j.$$

In mathematical language, equations (2) define a representation of the multiplicative *group*[3] of positive n-vectors, defined by (3), as a group (2) of linear transformations of the space of vectors Q.

61. Unit-free relations

Two further assumptions were used in deriving the formula $v = C\sqrt{g\lambda}$ of Example 1. These can be formulated, in the general language of definitions I and II above, as follows.

III. There is a functional relation of the form

$$(4) \qquad \phi(Q_0, \cdots, Q_r) = 0,$$

where ϕ is a single-valued function. (In Example 1, $r = 2$ and $\phi = v - f(\lambda, g)$.)

IV. The relation (4) is independent of the choice of fundamental units.

Dimensional analysis thus deals in general with functional relations $\phi(Q_0, Q_1, \cdots, Q_r) = 0$ between *positive* variables Q_i, which transform by formula (2) under the (n-parameter, commutative) group of T_α. More precisely, it treats unit-free functional relations, defined as follows.

Definition. A functional relation $\phi(Q_0, Q_1, \cdots, Q_r) = 0$ between dimensionally homogeneous variables is *unit-free* if and only if

$$(5) \qquad \phi(Q_0, \cdots, Q_r) = 0 \quad \text{implies} \quad \phi(T_\alpha(Q_0), \cdots, T_\alpha(Q_r)) = 0$$

for every change T_α of fundamental units.

---

[3] By a *group* is meant a set closed under associative multiplication, each of whose elements has a multiplicative inverse. See [42], Ch. VI, for details.

Using (3b) it follows, conversely, that

$$(5')\qquad \phi(T_\alpha(Q_0), \cdots, T_\alpha(Q_r)) = 0 \quad \text{implies} \quad \phi(Q_r, \cdots, Q_r) = 0$$

for every T_α—but we will avoid this subtlety by assuming both (5) and (5'). In other words, we define $\phi(Q_0, Q_1, \cdots, Q_r) = 0$ to be unit-free if and only if, in the hyperoctant defined by positive[4] vectors $Q = (Q_0, Q_1, \cdots, Q_r)$, the *locus* defined by the equation is *invariant under the group* (2).

The idea that $\phi = 0$ refers to a locus, and that there is no restriction on the form of the function ϕ, should be stressed. The distinction is illustrated by the following simple example.

Example 2. Consider particles accelerated from rest with time-independent acceleration a. Denote distance by s, time by t, and velocity by v.

For such a system the familiar dimensionally homogeneous (hence, unit-free) relations $v = at$ and $v^2 = 2as$ always hold. But now, the locus defined by $v = at$ in the positive octant of (v, a, t)-space is the same as that defined by the bizarre relation

$$(6)\qquad \sqrt[3]{(v - at)(a + v)} + \sqrt[5]{t(v - at)} = 0.$$

Hence (6) is *unit-free*, though not dimensionally homogeneous.

Though this example is artificial, it will help to bring out the difference between Buckingham's formal proof (§64) of the Pi Theorem and the more general geometrical proof of Vaschy, which will be given in §63. But before proving the Pi Theorem in general, we first consider the special case $r = n$, of unit-free relations

$$(7)\qquad Q_0 = f(Q_1, \cdots, Q_n)$$

involving just one more quantity Q_j $(j = 0, \cdots, n)$ than fundamental unit $(i = 1, \cdots, n)$. We will assume that f is a single-valued function, and also that the quantities $Q_1, \cdots, Q_n$ involve all n fundamental units—i.e. that the $(n + 1) \times n$ matrix $\|b_{jk}\|$ in (2) has rank n. An equivalent condition is, that the square submatrix B corresponding to $j = 1, \ldots, n$ should be *non-singular*.[5]

As already stated below Eq. (1), Example 1 refers to this case, with $n = r = 2$.

Theorem 1. Any unit-free relation (7) which involves n fundamental units is equivalent to

$$(8)\qquad Q_0 = CQ_1{}^{x_1} \cdots Q_n{}^{x_n},$$

[4] Extensions can be made to non-positive vectors, but only at the expense of more complicated statements.

[5] An equivalent condition is $|B| = \det \|b_{ij}\| \neq 0$. (See [42], p. 304.)

where $C=f(1,\cdots,1)$ and the x_i are determined by

$$(8') \qquad b_{0i} = b_{1i}x_1 + \cdots + b_{ni}x_n, \quad [i=1,\cdots,n].$$

Proof. Since the Q_i are positive and B is non-singular, we can find α such that $T_\alpha(1)=Q$, where $1=(1,\cdots,1)$; let $C=f(1,\cdots,1)$. Hence, by (5)–(5') applied to $C-f(1,\cdots,1)=0$,

$$T_\alpha(C)-f(Q_1,\cdots,Q_n) = 0.$$

Applying (2) to the first term, and transposing,

$$(9) \qquad f(Q_1,\cdots,Q_n) = C\alpha_1^{b_{01}}\cdots\alpha_n^{b_{0n}}.$$

On the other hand, since B is non-singular, (8') has a unique solution $x=(x_1,\cdots,x_n)$. For this x,

$$Q_j^{x_j} = T_\alpha(1)^{x_j} = (\alpha_1^{b_{j1}}\cdots\alpha_n^{b_{jn}})^{x_j}.$$

Elementary exponential algebra thus gives

$$\prod_j Q_j^{x_j} = \prod_{j,k} \alpha_k^{b_{jk}x_j} = \prod_k \alpha_k^{\Sigma\, b_{jk}x_j} = \prod_k \alpha_k^{b_{0k}}.$$

Substituting into the right side of (9), we get (8).

The following familiar examples further illustrate the meaning of Theorem 1.

Example 3. Assume that the fluid resistance D for a rigid object of a given shape and aspect is *inertial*, in the sense that it is determined by the fluid density ρ, speed v, and object diameter d. Then (8), with $x_1=x$, $x_2=y$, $x_3=z$, is equivalent to

$$MLT^{-2} = (ML^{-3})^x(LT^{-1})^yL^z,$$

so that equations (8') reduce to

$$1 = x, \qquad 1 = -3x+y+z, \qquad -2 = -y,$$

whence $x=1$, $y=z=2$. Hence if the relation is unit-free, then $D=K_D\rho v^2 d^2$, where K_D is a constant. (Actually, $K_D=\pi C_D/8$, which is called the ballistic drag coefficient, varies slowly.)

Example 4. If ρ, v, d, and the fluid viscosity μ determine D by a unit-free functional relation, and if inertia is negligible (Stokes' "creeping flows"), then $D=K^*\mu v d$ for another constant K^*, by a similar dimensional calculus.

62. Reynolds and Mach numbers

In Theorem 1 the number n of fundamental units was equal to the number r of variables involved in a unit-free relation $Q_0=f(Q_1,\cdots,Q_r)$.

When $r=n+1$, manipulations similar to those of Theorem 1 lead to formulas involving useful dimensionless parameters.

Theorem 1′. Any unit-free relation $Q_0=f(Q_1, \ldots, Q_r)$ which involves $r-1$ fundamental units, can be written in the form

$$Q_0 = C(\Pi)Q_1{}^{x_1} \cdots Q_r{}^{x_r},$$

where $\Pi = Q_1{}^{a_1} \ldots Q_r{}^{a_r}$ is a dimensionless power product of $Q_1, \ldots, Q_r$, and $Q_0 = Q_1{}^{-x_1} \ldots Q_r{}^{-x_r} = \Pi_0$ is also dimensionless.

The preceding result, which is a special case of the Pi Theorem to be proved below, will now be illustrated by two important examples in fluid mechanics.

Example 5. Assume that $D=f(\rho, v, d, \mu)$ is a unit-free function of ρ, v, d and μ, under all changes (1) of units of length, time, and mass. The dimensionless quantities $K_D = D/\rho v^2 d^2$ and $R = \rho v d/\mu$ (Reynolds number) are invariant under such changes. By one such change we can, however,[6] reduce ρ, v, d simultaneously to 1; this will carry μ into $\mu/\rho v d = 1/R$. Therefore

$$(10) \qquad D = K_D(R)\rho v^2 d^2, \quad \text{where} \quad K_D(R) = f\left(1, 1, 1, \frac{1}{R}\right).$$

Example 6. Assume that D is similarly determined by ρ, v, d, and the compressibility $-d(1/\rho)/dp = d\rho/\rho^2 dp$ of the undisturbed fluid. These yield the dimensionless power products $D/\rho v^2 d^2 = K_D$ and $v^2 d\rho/dp$ (of dimensions $(LT^{-1})^2(ML^{-3})(MLT^{-2}L^{-2})^{-1}=1$). A more physical interpretation of $v^2 d\rho/dp$ is obtained if we recall that $dp/d\rho = c^2$, where c is the speed of sound in the fluid. Repeating the argument of Example 5, we then get similarly

$$(11) \qquad K_D = f(M^2), \quad \text{where} \quad M = \frac{v}{c} \text{ is the Mach number.}[7]$$

Formula (10) can also follow from Theorem 2 of Chapter II, provided one assumes that the Navier-Stokes equations accurately determine the fluid motion; cf. §71. Formula (11) can be deduced similarly from the Euler-Lagrange equations; cf. §73.

Before proving the Pi Theorem, one more important application of dimensional analysis will now be given.

Example 7. Let there be a *stationary* distribution of turbulent energy E per unit mass among different eddy sizes λ, so that $dE = E'(\lambda)d\lambda$. Assume that this distribution is determined by an *inertial* mechanism of turbulent

---

[6] This is essentially Vaschy's argument; see also D. Riabouchinsky, *L'Aerophile*, Sept. 1911.

[7] Called the Sarrau number in France. The term "Mach number" was proposed by J. Ackeret, *Schweiz. Bauzeitung, 94* (1929), p. 179.

energy transfer to smaller eddy sizes λ. Then the rate of energy transfer, per-unit-mass and time, to smaller eddies, clearly has dimensions $V^2/T = L^2/T^3$; hence it is multiplied by α^2/γ^3 under any change of scale $L \to \alpha L$, $T \to \gamma T$. Moreover, to maintain $E'(\lambda)$ unchanged, this rate must be independent of λ. Hence the mean *time* $T(\lambda)$ required to break up eddies of size λ into smaller eddies must be proportional to $\lambda^{2/3}$: T *has the dimensions* $L^{2/3}$ *under change of scale*. Now consider the *frequency* spectrum of energy: $dE = F(k)dk$, where $k = 2\pi/\lambda$ is the wave number. Since dE has dimensions $V^2 = L^2/T^2$, while k and $dk = 2\pi d\lambda/\lambda^2$ have dimensions $1/L$, $F(k)$ has dimensions L^3/T^2, or $L^{5/3}$, or $k^{-5/3}$. In conclusion, dimensional analysis gives Kolmogoroff's formula for *turbulent energy distribution*: $F(k) \propto k^{-5/3}$.

Kolmogoroff's formula involves the well-known paradox of infinite turbulent total energy density per-unit-volume for large eddies; I shall not discuss the rationalization of this paradox.

63. Pi Theorem

Instead of giving more examples of special cases,[8] I shall proceed directly to the general Pi Theorem of Vaschy and Buckingham, which may be formulated as follows:

Theorem 2. Let positive variables $Q_1, \cdots, Q_r$ transform by (2) under all changes (1) in fundamental units $q_1, \cdots, q_n$. Let $m \leq n$ be the "rank" of the matrix $\|b_{ik}\|$ defined by (2). Then any unit-free relation of the form

$$(12) \qquad f(Q_1, \cdots, Q_r) = 0$$

is equivalent to a condition of the form

$$(13) \qquad \phi(\Pi_1, \cdots, \Pi_{r-m}) = 0$$

for suitable dimensionless power-products $\Pi_1, \cdots, \Pi_{r-m}$ of the Q_i.

Explanation. The first sentence restates Assumptions I and II of §60. Assumptions III and IV are generalized in (12).

Proof. By matrix theory $\|b_{ik}\|$ has a non-singular $m \times m$ minor.[9] By permuting the Q_j and q_i, we can suppose that this minor involves $Q_1, \cdots, Q_m$ and $q_1, \cdots, q_m$. (Thus, in the usual language of physics, the other fundamental units are not involved independently.) Then any

---

[8] Numerous other examples are given in Bridgman [43], Chs. I, VI; in Sedov [56]; in Langhaar [51]; in Porter [53]; and in B. A. Robertson, *Gen. Elec. Review, 33* (1930), 207. See also Rayleigh, *Phil. Mag., 34* (1892), 59; and *8* (1905), 66; also *Nature, 95* (1915), 66.

[9] For the properties of matrices assumed here, see for instance [42], Ch. X, esp. p. 306. A non-singular $m \times m$ minor is simply a square submatrix whose determinant is not zero.

vector $b_j = (b_{j1}, \cdots, b_{jn})$ with $j > m$ is a linear combination $b_j = c_{j1}b_1 + \cdots + c_{jm}b_m$ of the vectors $b_1, \cdots, b_m$.

We now define $r - m$ new *dimensionless* variables Π_i by

$$\Pi_i = Q_{m+i}Q_1^{-c_{j1}} \cdots Q_m^{-c_{jm}}.$$

We define a new function g by

$$(14) \qquad g(Q_1, \cdots, Q_m, \Pi_1, \cdots, \Pi_{r-m}) = f(Q_1, \cdots, Q_r);$$

in (14), clearly $Q_j = \Pi_{j-m}Q_1^{c_{j1}} \cdots Q_m^{c_{jm}}$, for $j > m$.

In the "octant" $Q_1 > 0, \cdots, Q_r > 0$, the transformation (2) of independent variables is one-one *in the large*.[10] Hence $f = 0$ is equivalent to (i.e. defines the same locus as) $g = 0$, and $g = 0$ is consequently also unit-free. But since the minor $\|b_{ij}\|$ with $i, j = 1, \cdots, m$ is non-singular, the simultaneous linear equations $b_{i1} \log \alpha_1 + \cdots + b_{im} \log \alpha_m = \log Q_i$ can be solved for arbitrary positive $Q_1, \cdots, Q_m$ by suitable choice of positive $\alpha_1, \ldots, \alpha_m$. Since the relation $g = 0$ is unit-free, the locus is the same for all $Q_1, \cdots, Q_m$; hence (12) is equivalent, for example, to

$$\phi(\Pi^1, \cdots, \Pi_{r-m}) = g(1, \cdots, 1; \Pi_1, \cdots, \Pi_{r-m}) = 0.$$

This completes the proof.

Historical Note. There is some controversy over the origins of the Pi Theorem. Vaschy[11] stated the result in 1892, but without formulating his assumptions. He indicated the method used above, but his logic was so cryptic that nobody reproduced his demonstration. Buckingham ([45], [46]) gave in 1914 the first proof of the Pi Theorem, but for the special case that f could be expanded in a Maclaurin series; until recently this was the only accepted proof.[12] Recently, Riabouchinsky and A. Martinot-Lagarde [52] have clarified the ideas of Vaschy so as to obtain a much more general proof. The proof given here is intended to bring out more clearly the restriction to positive α_i and Q_j, and the theorems on matrices assumed.[13]

[10] In case the c_{jm} are integers, this can be generalized to other octants. The behavior on the hyperplanes $Q_i = 0$ is more difficult, and has been discussed by D. Riabouchinsky, *Comptes Rendus*, 217 (1943), 220–223.

[11] A. Vaschy, *Annales Télégraphiques*, 19 (1892), 25–28. The ideas of Riabouchinsky were evolved in a series of papers (*L'Aerophile*, Sept. 1911; *Comptes Rendus*, 217 (1943), 205–208, and 225 (1947), 837–839).

[12] In fact, Bridgman [43], p. 16, raises the question of whether more general functions can be considered. A dimensionless function for which expansion in Maclaurin series is impossible is the $\alpha(\beta, M, \gamma)$ in the Taylor-Maccoll determination of §85. See also the Ferri Paradox of §16.

[13] See also H. L. Langhaar [51] and refs. given there.

64. Discussion of proof

If one assumes that f can be expanded in a Maclaurin series, then one can give an alternative algebraic proof of the Pi Theorem which is perhaps easier to comprehend. I will present this proof here, together with some related results, to clarify further the concept of *dimensional homogeneity*. First, note the following obvious consequences of Euler's characterization of homogeneous functions.

Lemma 1. For a function $f(Q)$ of positive quantities $Q_1, \cdots, Q_m$, Euler's homogeneity relations

$$(15) \qquad \frac{\partial f}{\partial Q_j} = \frac{\lambda_j f}{Q_j}, \qquad \lambda_j \text{ real constants,}$$

are equivalent to the statement that

$$(15') \qquad f(Q) = C Q_1^{\lambda_1} \cdots Q_m^{\lambda_m}, \quad C = f(1, \cdots, 1).$$

If the Q_j are dimensionally homogeneous, as in (2), then so is f, and its dimensions in q_k are $\lambda_1 b_{1k} + \cdots + \lambda_m b_{mk} = \Lambda_k$.

To deal with such functions, we make a

Definition. A function $f(Q)$ satisfying (15') will be called a *Q-monomial.* A finite sum

$$(16) \qquad \phi(Q) = f_1(Q) + \cdots + f_r(Q)$$

of Q-monomials will be called a *Q-polynomial.*

Lemma 2. If all terms $f(Q)$ of $\phi(Q)$ have the same dimensions Λ_k in every q_k, then ϕ is dimensionally homogeneous. In fact, substituting from (2),

$$(17) \qquad \phi(T_\alpha(Q)) = \phi(Q, \alpha) = \alpha_1^{\Lambda_1} \cdots \alpha_n^{\Lambda_n} \phi(Q).$$

The simple example $f_1 = Q_1$, $f_2 = Q_2$, $f_3 = -Q_1$ shows that the converse is not true unless ϕ has been reduced to normal form. We define a Q-polynomial to be *formally homogeneous* if all its terms f_i have the same dimension-vector $\Lambda = (\Lambda_1, \cdots, \Lambda_n)$. Evidently if ϕ is formally homogeneous, then the relation $\phi = 0$ is unit-free in the sense (5). Moreover, it is equivalent to the dimensionless relation $1 + (f_2/f_1) + \cdots + (f_r/f_1) = 0$, which trivially proves the Pi Theorem for Q-polynomials.

Many equations coming from physics are formally homogeneous, like those in Examples 1 and 3–4 above. It has even been asserted (though incorrectly, see §65) that true physical equations *must* be homogeneous, and it is true that the criterion of dimensional homogeneity often gives a convenient formal check for vaguely remembered physical equations. However, the precise facts are fairly subtle; some of the fine distinctions involved are best shown by example. Accordingly, we reconsider Example 2 of §61.

If the Pi Theorem is applied to the unit-free relation (6), with v substituted for Q_1, and t for q_2, we get the relation

$$\sqrt[3]{(1-\Pi)(\Pi+1)} + \sqrt[5]{1-\Pi} = 0,$$

after some manipulation. This relation, unlike (6), is not only unit-free but dimensionally homogeneous, since all its terms have dimension zero in all fundamental quantities. By way of contrast, Buckingham's proof of the Pi Theorem would not apply to (6).

Following Bridgman,[14] we can also consider the polynomial equation

(18) $\phi(s, v, a, t) = v + v^2 - 2as - at = 0.$

Like (6), (18) is satisfied in Example 2; moreover, ϕ is a Q-polynomial. Equation (18) is *not* unit-free in the sense (5), and ϕ is not formally homogeneous: the substitution $s \to \alpha s$, $t \to \beta t$ transforms (18) into

(18') $\dfrac{\alpha}{\beta}(v - at) + \left(\dfrac{\alpha}{\beta}\right)^2 (v^2 - as) = 0.$

Since (18) is true in any system of fundamental units ("true in all units", though not "unit-free"), (18') is an identity in α and β. Hence, (18) implies both $v = at$ and $v^2 = as$. This argument can be generalized as follows.

Theorem 3. Let $\phi(Q)$ be a Q-polynomial, and let the relation $\phi(Q) = 0$ be true in all units. Then $\phi(Q) = 0$ is equivalent to a set of formally homogeneous equations.

The proof follows from a formal inspection of the identity

$$\phi(T_\alpha(Q)) = \sum \alpha_1{}^{\Lambda_{i1}} \cdots \alpha_n{}^{\Lambda_{in}} \phi_i(Q),$$

where the $\phi_i(Q)$ are the constituents of ϕ having different dimensions $\Lambda_i = (\Lambda_{i1}, \cdots, \Lambda_{in})$. The argument, applied to Maclaurin series, gives Buckingham's proof of the Pi Theorem; it would seem to be equally applicable to Laurent or real Dirichlet series, within their domains of convergence.

Finally, we recall Bridgman's property[15] of "absolute invariance of relative magnitude." A function of the variables $q_1, \cdots, q_n$ is said by Bridgman to have this property when it satisfies the functional equation

(19) $\dfrac{f(q_1', \cdots, q_n')}{f(q_1, \cdots, q_n)} = \dfrac{f(\alpha_1 q_1', \cdots, \alpha_n q_n')}{f(\alpha_1 q_1, \cdots, \alpha_n q_n)}$

for all positive α_1, q_i and q_i' [$i = 1, \cdots, n$]. The result follows.

---

[14] [43], p. 42. His discussion of (18) led him to propose a "vector calculus" of relations.

[15] See Bridgman [38], p. 21. Equation (19) evidently states that *ratios* of homogeneous quantities are invariant under changes of the fundamental units. The assumption that the q_i be positive, though not stated, is also required in Bridgman's proof, since he forms $\int dq_i/q_i$. And indeed, negative α_i usually have no physical meaning. The extension from differentiable to continuous functions is due to A. Martinot-Lagarde, *Comptes Rendus, 223* (1946), 136–137.

Theorem 4. Let $Q = f(q_1, \cdots, q_n)$ be a positive quantity whose numerical magnitude varies continuously with the fundamental units q_i, and which satisfies (19), so that ratios of numerical magnitudes are invariant under changes in the fundamental units. Then Q must satisfy (2).

Proof. Let $q_1' = \sigma$, $\alpha_1 = \sigma^m$, and all other variables 1. After transposition, (19) becomes

$$\lambda f(\sigma^m, 1, \cdots, 1) = f(\sigma^{m+1}, 1, \cdots, 1),$$

where $\lambda = f(\sigma, 1, \cdots, 1)/f(1, 1, \cdots, 1)$. By induction on m we get, for all positive and negative integers m:

$$(19') \qquad f(\sigma^m, 1, \cdots, 1) = \lambda^m f(1, 1, \cdots, 1).$$

Hence, setting $q_1 = 2^{m/n}$, $\sigma = \sqrt[n]{2}$, and $f(1, \cdots, 1) = C$, we get

$$f(q_1, 1, \cdots, 1) = C\lambda^m, \quad \text{and} \quad f(2, 1, \cdots, 1) = C\lambda^n.$$

Defining $a = \log_2[f(2, 1, \cdots, 1)/f(1, 1, \cdots, 1)] = n \log_2 \lambda$, so that $\lambda^m = 2^{ma/n}$, we get by substitution,

$$(20) \qquad f(2^{m/n}, 1, \cdots, 1) = C2^{ma/n} = C(2^{m/n})^a.$$

But the powers of 2 with rational exponents are an everywhere-dense set of positive real numbers. Hence, if f is continuous—in fact, unless f is non-measurable and everywhere discontinuous[16]—$f(q_1, 1, \cdots, 1) = Cq_1^a$ holds for all positive q_1.

Repeating the argument for the other indices, we get the conclusion of the theorem.

65. Are physical laws unit-free?

In §61, the property of being unit-free was treated as a mathematical hypothesis. Its physical applicability has also been hotly debated. Thus, some writers have loosely interpreted the plausible principle that "all units of measurement[17] are valid," to imply that all such units must give rise to the same universal physical laws. Thus, Tolman[18] asserted in 1914 that "the fundamental entities of which the physical universe is constructed are such that from them a miniature universe could be constructed exactly similar . . . to the present universe."

[16] For everywhere-discontinuous counterexamples, see G. Hamel, *Math. Annalen*, *60* (1905), 459–462. By a logarithmic transformation (19) reduces to a well-known additive functional equation. It is easy to show that, by (19), if f is discontinuous at one point, it must be everywhere discontinuous. For the nonmeasurability of f, see S. Banach, *Théorie des Opérations Linéaires*, Warsaw, 1933, p. 23.

[17] For a discussion of the meaning of a scale of measurement, see Norman Campbell, *Measurement and Calculation*, 1928. Strictly, one should distinguish between enumerative scales, ordinal scales, scales in which equality of difference can be observed, and true linear scales with zero.

[18] R. C. Tolman, *Phys. Rev.*, *3* (1914), 244–55.

That this implication is not logically necessary is easily seen if one recalls that, in a sense, all coordinate systems for space-time are equally possible. But a geocentric system, such as used in Ptolemaic astronomy, does not give rise to the same physical laws as a heliocentric system.

Moreover the interpretation is not true, even for the units of length, mass, and time in mechanics.[19] Indeed, the main point of special relativity is that the laws of mechanics, which include the basic force law

$$(21) \qquad F = \frac{d(m_0 u/\sqrt{1-u^2/c^2})}{dt},$$

are *not* unit-free for length and time independently, but involve the speed of light.

In quantum mechanics the intervention of Planck's constant h through the de Broglie formula $\lambda = h/mv$ for the wave-length of a particle, and the photoelectric equation $E = h\nu$, shows even more obviously that not all physical laws are dimensionally homogeneous. Here h is a universal constant, of the dimensions ML^2/T of action (energy × time). Another dimensional constant γ is involved in Newton's universal law of gravitational attraction[20] $F = \gamma mm'/r^2$; still others in the diameter of every ultimate particle, etc. We are thus impelled irresistibly to the conclusion that *there are no known "fundamental units" with respect to which all known physical laws are unit-free.*[21]

Indeed, the decision to call certain units "fundamental" (or primary) and others "derived" (or secondary) is one of convention and not of physical necessity. Thus it is sometimes convenient to treat force as independent of mass, length, and time.[22]

Material constants. The preceding statements are subject to an important reservation. The universality of "universal constants" may not be absolute. Thus, before Newton's time g must have seemed like a universal constant (this idea is still inherent in the use of engineering units). Again, the velocity $c = 1/\sqrt{K\mu}$ of electromagnetic waves was not treated by Maxwell[23] as a universal constant, but was regarded as

[19] One can, of course, rescue Assumption IV by the sophistry of defining a "fundamental unit" in a theory to be one for which Assumption IV holds, this may even be useful.

[20] In fact, according to Eddington (*Lond. Phys. Soc.*, 1918, p. 91), Nature practically dictates choosing units of length, time, and mass in a unique fashion, so that $\gamma = c = h = 1$. The limited laws of "gravitational similitude" have been discussed by Labocetta, *Elettrotecnica*, 1932, p. 1629.

[21] For further examples see Bridgman [43], p. 103.

[22] See Bridgman [43], p. 65. In thermodynamics we have the remarkable paradox of Riabouchinsky (ibid., p. 10); also the Stefan-Boltzmann radiation law $E/\text{Area} = KT^4$. The artificiality of the concept of a "fundamental unit" appears even more clearly with electromagnetic units; see J. Jeans, *Electricity and Magnetism*, Cambridge Univ. Press, 1941, pp. 14–15; also Brylinski, *Comptes Rendus, 215* (1942), 104.

[23] *Theory of Electricity and Magnetism*, Oxford, 1881, Arts. 784–7.

depending on the dielectric constant K and magnetic permeability of the substance under discussion. It would seem premature to dismiss as without foundation efforts such as those of Tolman[18] and Eddington[24] to deduce relations between universal constants from metaphysical principles.

66. Inspectional analysis

Though one cannot say the same for relativistic or quantum mechanics, *one can make arbitrary changes* (1) *in the scales of length, mass, and time in Newtonian continuum mechanics.* And it seems certain that the laws of Newtonian mechanics very nearly determine the behavior of real fluids under ordinary circumstances. Though such changes of scale may drastically alter material properties like density and viscosity, real fluids have such a wide range of densities and viscosity that the effect is not ordinarily noticeable.

The principle italicized above can not only be demonstrated through model experiments; it can also be deduced theoretically from the fundamental equations of fluid mechanics. The deduction rests on a simple meta-mathematical principle: *if a set of mathematical equations is invariant under a group, then the same is true of all consequences of these equations.*

As applied to the scalar transformations (1), this was in fact the method used by Fourier, Stokes, and other pioneer users of dimensional analysis, to test the validity of their arguments. It was clearly in Rayleigh's mind when he referred to "similitude," and its advantages are also recognized by Bridgman[25] who writes, "The advantage of (dimensional analysis) is that it is rapid but . . . it does not give as complete information as might be obtained by . . . a detailed analysis," and "dimensional analysis physically is not so instructive as the similarity condition."

The idea of "dynamic similitude" is commonly defined for fluid motions as follows.

Definition. Two fluid motions Φ and Φ' are called *dynamically similar* if they can be described by coordinate systems[26] which are related by transformation of space-time-mass, of the form

$$(22) \qquad x_i' = \alpha x_i, \qquad t' = \beta t, \qquad m' = \gamma m.$$

---

[24] A. Eddington, *Relativity Theory of Protons and Electrons*, Cambridge Univ. Press, 1935. See also [43], Ch. VIII, and E. T. Whittaker, *Space and Spirit*, Edinburgh, 1946.

[25] Rayleigh, *Phil. Mag.*, *34* (1892), p. 52, and *8* (1905), p. 66; also *Nature*, *95* (1915), p. 66; Bridgman ([43], p. 17). See also L. Schiffer, *ZaMM*, *24* (1944), 289–93.

[26] Actually we mean "Newtonian" coordinate systems, in which Newton's laws of motion are valid. Since the assumption that these exist begs the question of the possibility of dynamic similitude, there is some circularity in this definition.

It is easy to test the equations of a hydrodynamical theory for invariance under transformations of the form (22). Essentially, this was just what was done in proving Theorem 2 of §21. The most important equations to test were the Navier-Stokes equations for an *incompressible viscous fluid*,

$$(23) \qquad \frac{Du_i}{Dt} + \frac{1}{\rho}\frac{\partial p}{\partial x_i} = g_i + \nu \nabla^2 u_i, \quad i = 1, 2, 3.$$

The conclusion was that all equations were invariant in the absence of a free surface, *provided* ν and p/ρ (hydrostatic pressure being allowed for by Theorem 1 of §21) were scaled so as to preserve the Reynolds number vd/ν of Example 5, §62. (For the role of "material constants" like ν and ρ, see §65; their role is similar to that of K and μ in electromagnetic theory, where $c = c_0/\sqrt{K\mu}$.)

Various analogs to Theorem 2 of §21 will be proved in §§70–73, applicable to compressible nonviscous flow, incompressible flows with free surfaces, cavitation, etc. But first, the concept of inspectional analysis will be discussed in general, so as to bring out more clearly its relation to traditional dimensional analysis.

67. Relation to groups

Clearly, one can test mathematical equations for invariance under many transformations, besides the "changes of scale" described by (22). For example, the equations of physics are *all* invariant under translations and rotations of axes—a well-known principle which is essential for the mathematical treatment of most physical problems. In special cases one can take advantage of invariance under conformal or affine transformations (see §74).

In general, *inspectional analysis is applicable to any group of transformations.*[27] By a group of transformations I mean,[3] of course, a set of transformations which contains the identity, the inverse of any one, and the product of any two of its members.

This assertion is based on the logical axiom, discussed in §1, (C), and §26, that *if the hypotheses of a theory are invariant under a group G, then so are its conclusions.*[28] Conversely, *the set of all one-one transformations leaving any set of equations forms a group.*

After the "dimensional group" of transformations of the form (22), the

---

[27] This point was made by Mrs. Ehrenfest [49], p. 261; but as she gave no applications, it has been generally ignored. For elementary facts about groups, see [42], Chap. VI; for orthogonal matrices, see Ch. VIII.

[28] This principle, though widely assumed by mathematicians and physicists, is seldom explicitly formulated. See, however, G. Bouligand, *Théorie Générale des Groupes*, Paris, 1935, p. 3.

most important group for mechanics is the ten-parameter *Galilei-Newton* group. This group is generated by the three-parameter subgroup S of space-translations

$$(24) \qquad x_i' = x_i + c_i \quad [i = 1, 2, 3];$$

the one-parameter subgroup T of time-translations

$$(25) \qquad t' = t + c;$$

the three-parameter subgroup R of rigid rotations

$$(26) \qquad x_i' = \sum a_{ik} x_k,$$

where $\|a_{ik}\|$ is the most general 3×3 orthogonal matrix; and the three-parameter subgroup M of changes to *moving axes* being translated with constant velocity

$$(27) \qquad x_i' = x_i - b_i t.$$

Now it is easy to verify that Newton's three laws of motion are invariant under (24)–(27), and that these transformations leave unchanged the definitions of such material constants as density, viscosity, etc. (mass being assumed the same). Hence, theoretical *Newtonian mechanics is invariant under the Galilei-Newton group*, as well as under the group of transformations (22) of dynamic similitude. Experimentally, this principle has been verified in many different ways with very great precision, except at speeds comparable with that of light.[29]

68. Theory of modeling

We have indicated two basic advantages of inspectional analysis: it enables one to justify Assumption IV of dimensional analysis by permitting one to *test* the equations defining a given boundary-value problem for invariance under (1); and it enables one to consider "similarities" not of the simple form (1). Inspectional analysis has a third advantage: in principle, it suggests a rational method for testing the validity of Assumption III.

Although, as we have seen, Assumptions I, II, and IV seem generally acceptable in fluid mechanics, it is otherwise with Assumption III. Moreover, dimensional analysis provides no basis for deciding a priori when variables $Q_1, \cdots, Q_n$ do determine Q_0 with sufficient accuracy. Thus Bridgman[30] says flatly that this crucial question "cannot be decided by the philosopher in his armchair," but only on the basis of considerable physical experience. I shall give many experimental instances of this difficulty.

[29] The seeming Paradox of Dubuat is not a counterexample; see §28.
[30] [43], pp. 13–14; see also ibid., p. 50.

To test the validity of Assumption III by inspectional analysis, one can proceed in principle as follows. Let it be known that a certain fluid flow can be approximately predicted by solving an appropriate *boundary-value problem*, in the sense of §1. One then simply tests the differential equations and boundary conditions involved for invariance under transformations of a specified group (say (22)). If they are invariant and the boundary-value problem is "well-set," then Assumption III is valid.

Inspectional analysis thus has the advantage of fitting into the general pattern of rational hydrodynamics. The main limitation on its rigorous applicability comes from the fact, already brought out in Chs. I–II, that so few problems of real hydrodynamics have yet been reduced to demonstrably "well-set" boundary-value problems.

69. Partial inspectional analysis

I shall now give an example which illustrates both inspectional analysis and the care with which it must be applied.

Consider the Navier-Stokes equations (23) for an incompressible viscous fluid. Following Ruark [55], they can be reduced to dimensionless form as follows.

Let V, L, and P be respectively a characteristic velocity, length, and pressure in the model, presumably controlled on the boundary of the fluid flow. If we multiply (23) by L/V^2 to make it dimensionless, and introduce the dimensionless variables $u_i' = u_i/V$, $t' = Vt/L$, $x_i' = x_i/L$, $p' = p/P$, and the dimensionless constants $R = VL/\nu$, $F = V^2/Lg$, and $Q^* = 2P/\rho V^2$, we get

$$(28) \qquad \frac{Du_i'}{Dt'} = \frac{1}{R} \nabla'^2 u_i' + \frac{1}{F}\left(\frac{g_i}{g}\right) - \frac{Q^* \partial p'}{2 \partial x_i'},$$

where the g_i/g are the direction cosines of gravity.

There is a remarkable correlation between the dimensionless differential equation (28) and engineering experience; we can derive from it three of the most important rules of thumb involved in model work.[30a] Thus it suggests that if neither gravity, compressibility, nor cavitation are important, one should use models at the same Reynolds number R. If compressibility, cavitation, and viscosity are unimportant, one should use models at the same Froude number F. If compressibility and viscosity are unimportant, but gravity and cavitation effects are involved, one should preserve both F and the "cavitation number" Q^* (see §§72, 78).

It also gives the plausible suggestion that quantities are unimportant

---

[30a] See [11], vol. 2, Ch. I, for a similar derivation. The prescription is that "ratios of forces" must be preserved; neither Q^* nor differential equations are discussed explicitly.

precisely when corresponding coefficients in (28) are small, and so appears to give a complete theoretical basis for model work based on F, R, and Q^*.

Although the argument just given is suggestive and worth remembering, it suffers from the defect of considering only one of the three fundamental equations of fluid mechanics, the equation of motion. Thus it ignores the "equation of continuity"

$$(29) \qquad \frac{\partial \rho}{\partial t} + \sum \frac{\partial(\rho u_k)}{\partial x_k} = 0, \quad \text{or} \quad \operatorname{div} u + \frac{D(\log \rho)}{Dt} = 0,$$

and the "equation of state," which can be written

$$(30) \qquad \rho = f(p).$$

For this reason I shall call it "partial inspectional analysis," and shall designate the corresponding procedure, when "well-set" conditions completely determining a flow have been considered, "complete inspectional analysis."

70. Inertial modeling

In practice, considerations of experimental convenience and economy often dictate not only the fluid used (air or water), but also the *size* of models and the *speed* of flow. The use of small models to simulate full-scale behavior is ordinarily justified by an appeal to dimensional analysis, Specifically, one often assumes the approximate validity of the following

Principle of inertial modeling. Dimensionless quantities are preserved under all transformations of the form (22).

Thus, if L is a representative length and V a representative velocity, then $V^{-1}u(L^{-1}x)$ is assumed to be invariant under (22). As a corollary, one can deduce the full-scale $v(y)$ from the model $u(x)$ by setting

$$v(y) = \frac{V'}{V}\, u\!\left(\frac{Ly}{L'}\right),$$

where L', V' are representative full-scale length and velocity. Similarly, one assumes the invariance under (22) of the pressure coefficient $C_p = (p-p_a)/\tfrac{1}{2}\rho v^2$, where p_a is the ambient pressure. In a nonviscous fluid this implies the invariance of $C_D = D/\tfrac{1}{2}\rho V^2 A$, where D is the drag and A the cross-section area. Note that preservation of $p/\tfrac{1}{2}\rho v^2$ is however *not* assumed (see §72): dimensional analysis is not swallowed whole.

Actually, the method of inspectional analysis enables one to dispense with the assumptions of dimensional analysis altogether. In particular, one can *deduce* the principle of inertial modeling from the standard equations for incompressible nonviscous flow, in the absence of a free surface.

Thus, almost trivial calculations show that the substitutions (22), combined with $u' = \alpha u/\beta$, $\rho' = \gamma \rho/\alpha^3$, and $p' = (\gamma/\alpha\beta^2)p$, preserve Euler's equations of motion and the Equation of Continuity—as well as the condition $\nabla \times u = 0$ of irrotational flow. This proves

Theorem 5. The principle of inertial modeling is applicable to Euler's equations of motion, to the Equation of Continuity and irrotationality, and to Euler's boundary conditions at solid boundaries, in incompressible flow.

Since the above conditions define a well-set boundary-value problem (the Neumann problem, see §4) for steady flow, given p_a, there follows the

Corollary. If Euler's equations for irrotational incompressible flow are valid, measured C_D should be independent of size, speed of motion, and fluid density.

Actually, in view of the d'Alembert Paradox, this result is less interesting in itself than as an illustration of an important method. However, the argument applies equally well to Joukowsky flows (§8), to Kirchhoff "wakes,"[31] (§39), to Helmholtz-Brillouin flows (§47), and to Karman's theory of vortex streets (§56). The principle of inertial modeling also holds for Newton's primitive kinetic theory of air resistance, and for Euler's quasi-empirical formula expressing drag and lift as definite integrals.[32]

Theorem 5, as applied to acceleration from rest, also predicts that the added mass coefficient k, defined as the ratio[33]

$$k = \frac{\text{added mass}}{\text{mass of fluid displaced}},$$

will be determined by the *shape* of the body considered, independent of its size, rate of acceleration, or the fluid density.

Experimentally, the principle of inertial modeling is approximately valid over extensive "regimes," corresponding to large ranges of R. However, it breaks down abruptly when there is transition to alternating vortices and to boundary-layer turbulence (e.g. in the vicinities $1/R \simeq 0.02$ and 0.00005, as explained in §28).

71. Reynolds modeling

A much more satisfactory application of the method of inspectional analysis is to the equations for incompressible viscous fluids. In this case it yields, by Theorem 2 of §21.

Theorem 6. If the Navier-Stokes equations for an incompressible

---

[31] i.e. to cavities. Note that as the theories of Euler and Joukowsky flows are reversible, we can even have $\alpha < 0$ in (22).

[32] These theories are discussed at length in P. Painlevé, *Leçons sur la Résistance des Fluides*, Paris, 1930, and in Cranz [3], Ch. II.

[33] The *added mass* of a body in a fluid (Ch. VI) is the difference between the inertia which it has in the fluid and in vacuo.

viscous fluid, together with the hypotheses of incompressibility and no-slip, approximately determine a (statistically) time-independent fluid flow, then (8) holds.

Experimentally, equation (8) has received impressive experimental confirmation with the most diverse liquids and gases.[34] Thus (as shown in Fig. 8) the breakdown of Poiseuille flow in pipes comes at the same Reynolds number in air, water, and many other fluids. At Mach numbers below $M = 0.3$, drag coefficients of spheres and cylinders satisfy (8) with the same $K_D(R)$ in all fluids, and at all sizes and speeds. A corresponding result holds for the skin-friction of plates parallel to the wind.

In verifying these facts, two precautions must be taken; otherwise Reynolds modeling will not work. First, one must use models with similar *surface roughness*. The onset of turbulent flow and transition in the boundary layer from laminar to turbulent flow are greatly influenced by this factor. Thus near R_{crit}, the drag of a sphere can be greatly reduced by roughening it suitably.

Second, especially in wind-tunnels with closed circuits, the *turbulence of the free stream* must be the same.[35] It is found that the R_{crit} of spheres in wind-tunnels may vary by a factor of two, depending on the turbulence of the wind-tunnel. A practical solution to this problem will be described in §75.

It is not easy to reproduce large Reynolds numbers economically on a small scale, in high-speed flow. If one tries to use a given fluid (air or water) under atmospheric conditions, then any reduction in diameter must be compensated for by increasing the velocity in the same ratio. With air, one can reduce ν by using compressed air, to compensate for a reduction in linear scale; (cf. the end of §73 and §75). Unfortunately, no liquid is known whose ν is much less than that of water, though many have much greater ν. Hence wind-tunnels[36] provide the only known economical Reynolds models of phenomena of water flow.

72. Froude and cavitation modeling

Inspectional analysis can also be used to derive modeling laws for situations in which viscosity and compressibility are unimportant, but a

---

[34] See §25 and refs. given there; also [53], pp. 16–17. The theory is due to Stokes, [13], Vol. 3, p. 17. Since the turbulent behavior of liquids is dynamically similar to that of gases, it seems hardly possible that turbulence should be related to kinetic theory except indirectly, through viscosity. Similar studies of oils have been made by R. O. Boswall and J. C. Brierly, *Proc. Inst. Mech. Eng.*, *122* (1932), 423–569, for lubrication.

[35] Goldstein [4], p. 431; see also §28, where similar observations are made.

[36] For the substitution of wind-tunnels for water tunnels, see C. Keller, Escher Wyss News. (1940). Superfluid He (§20) seems unsuitable.

"free surface" at constant pressure is involved. In particular, such laws apply to gravity waves and cavitation phenomena in liquids. We have

Theorem 7. In a uniform gravity field with intensity g, Euler's equations of motion and boundary conditions at solid boundaries, together with irrotationality and the "free surface" condition $p = $ const. at a liquid-gas interface, are preserved by all transformations (22) which preserve the Froude number $F = V^2/gL$.

Proof. By Theorem 5, it suffices to consider the free surface condition $p = $ const.—i.e. the condition that ∇p be normal to the interface. To prove this we multiply (23) by L/V^2 as in deriving (28), getting the dimensionless equation

$$(31) \qquad \frac{Du_i'}{Dt'} = \frac{1}{F} \left(\frac{g_i}{g} \right) - \frac{\partial(p/\rho_0 V^2)}{\partial x_i'}.$$

Since g_i/g is the i-th direction cosine of the gravity field, and since $u(x, t)$ determines ∇p, we have dynamic similarity with proportional *differences* in the pressure coefficient $2p/\rho_0 V^2$ (though not usually proportional $2p/\rho_0 V^2$), provided we have equal F.

Indeed, the ambient pressure P is usually the local atmospheric pressure p_a in gravity wave modeling; in cavitation modeling the vapor pressure p_v must also be considered. This has only been clearly recognized since 1924, when Thoma[37] introduced his cavitation number

$$(32) \qquad Q = \frac{P - p_v}{\frac{1}{2}\rho V^2}.$$

Before then, cavitation was generally stated to depend on the homogeneous dimensionless parameter

$$(32') \qquad Q^* = \frac{p}{\frac{1}{2}\rho V^2}$$

of (28), which is directly suggested by ordinary dimensional analysis.[38]

Complete inspectional analysis, coupled with the assumption that cavitation occurs spontaneously when $p < p_v$, gives a rational basis for preferring (32). For, the preceding assumption amounts to postulating the discontinuous Equation of State of Ch. III, (14),

$$(33) \qquad \begin{aligned} \rho &= \rho_0 \quad \text{if} \quad p > p_v \quad \text{and} \\ \rho &= \rho_v, \quad p = p_v \quad \text{elsewhere.} \end{aligned}$$

[37] D. Thoma, Experimental Research in the Field of Water Power, *Trans. First World Power Conf.*, Vol. 2 (1924), 536–551; see also H. B. Taylor and L. F. Moody, *Mech. Engineering*, *44* (1922) 633–640. An explicit statement was given by H. Lerbs on p. 290 of *Hydromechanische Probleme des Schiffsantriebs*, Hamburg; see also H. E. Rossell and L. B. Chapman, *Principles of Naval Architecture*, Soc. Nav. Arch. Marine Eng., New York, 1947, Vol. 2, p. 177.

[38] See F. Lorain, *L'hélice Propulsive*, Paris, 1923, p. 129; E. Buckingham, *Jour. Am. Soc. Naval Eng.*, *48* (1936), 147–198; D. W. Taylor, *The Speed and Power of Ships*, 3d ed. (1943), p. 17.

Given P and p_v, a transformation of similitude (22) preserves (33) if and only if it preserves Q; the proof is similar to that of Theorem 7.

One can preserve both gravity *and* cavitation effects in models at $1:\alpha$ linear scales, by using $1:\sqrt{\alpha}$ velocity scales (same g), and controlling P so as to transform $P-p_v$ in the ratio $1:\alpha$. This "reduced pressure Froude modeling" is now widely used in model studies of marine propeller cavitation; in the model, vapor pressure may not be negligible.

73. Mach modeling

It has been known since the early experiments of Robin[39] (1742), that the drag of bullets is not proportional to the velocity squared; hence no form of inertial scaling is valid. In the notation of Example 3 of §61, K_D has a notable rise near the speed of sound. As a result, K_D has been commonly tabulated as a function of v.

It was recognized from the start, that this was due to the compression of the air, but the more rational tabulation of K_D as a function of the Mach number M dates since World War I. It follows logically that there should be a temperature effect on the range of projectiles, in addition to the effect of density already obvious from the definition of K_D. However, this was apparently first stated explicitly after the first World War.[40]

Beginning about 1935 with the advent of high-speed planes, the aerodynamicists have become interested in Mach modeling. Whereas wind-tunnels operating at 100 ft/sec can be used to simulate flight conditions up to 400 ft/sec, provided the "effective" Reynolds number is properly controlled, they give no idea at all of the compressibility effects which arise at higher speeds. Since 1935, therefore, the aerodynamicists and ballisticians have joined forces in the study of compressible flows.

The laws of Mach scaling were first derived by Langevin,[40] from an "inspectional analysis" of the equations for a *compressible nonviscous* gas neglecting gravity, in the case of a polytropic equation of state. I shall give a slight generalization of his results.

Going back to Theorem 5, we see that all transformations of similitude preserve the equation of continuity. Obviously also, a given equation of state (30) is preserved under any transformation which preserves p and ρ

---

[39] See Cranz [3], pp. 44–45, for an historical account.

[40] G. Darrieus, *Mem. Art. Française* (1922), p. 242; H. W. Hilliar, *Dept. Sci. Res. Exp. Report RE 142/19* (1919); Darrieus states that the variation in range due to this cause can be as much as 1%. The article by Langevin cited below comes immediately after that by Darrieus. Mach modeling was justified, on the basis of dimensional analysis, by Buckingham [46], pp. 275–278. For practical aspects, see R. H. Kent, *Mech. Eng.*, Sept. 1932.

at corresponding points. Hence, in particular, it is preserved under the transformations

(34) $x_i \to \alpha x_i, \quad t \to \alpha t; \qquad p, \rho, \boldsymbol{u}$ unchanged.

By inspection, the terms $Du_i/Dt = \sum u_k \partial u_i / \partial x_k + \partial u_i / \partial t$ and $\partial p / \rho \partial x_i$ in (23) are both multiplied by $1/\alpha$ under (34). We conclude

Theorem 8. The fundamental equations for compressible nonviscous flow are invariant under (34).

As shown in Chapter I, these fundamental equations do not define a well-set boundary-value problem. At the least, one must adjoin the Rankine-Hugoniot equations for shock waves (§14). However, since these equations can be derived from the equation of state and the laws of conservation of mass, momentum, and energy, and these laws are preserved under any transformation (34), the Rankine-Hugoniot equations are scaled also.

The scaling law (34) holds also in elasticity, plasticity, and in explosion dynamics;[41] it has been called the Law of Cranz. More generally, it holds whenever the *stress tensor is a function of strain only*, independent of the rate of strain—and whenever a fixed chemical energy per-unit-volume is released in conjunction with a particular state of strain, as postulated in the Chapman-Jouguet conditions ([2], §87). Curiously, it also holds in *relativistic* fluid mechanics.

Partial inspectional analysis has been thought by some to give a basis for Mach scaling. Let $c = \sqrt{dp/d\rho}$ denote the local speed of sound, and let C be the sound velocity of the free stream. Then if the forces of viscosity and gravity are neglected, and we write $M = V/C$, (23) becomes

(35) $$\frac{\partial u'_i}{\partial t'} + \sum \frac{u'_k \partial u'_i}{\partial x_k} + \frac{1}{M^2}\left(\frac{c}{C}\right)^2 \frac{\partial p}{\rho \partial x_i} = 0.$$

This suggests the plausible rule: *scale at the same Mach number*, which for a given free stream is equivalent to (34).

However, this rule is *false* in general, if we consider behavior in different gases (gases with different equations of state (30)), or even the same gas under different temperatures and pressures. Compare, for example, dynamically similar homentropic flows of a gas, in which the free stream conditions correspond to two points along the same adiabatic. By (22), $\boldsymbol{u}$ and ρ are each everywhere multiplied by constant factors. Hence by (35), grad p is multiplied by a constant factor $a(\alpha)$, where α is the ratio of free stream densities. Hence if $F(\rho) = p(\rho) - p_f$ (p_f=free stream pressure), $F(\alpha\rho)/F(\rho) = a(\alpha)$ is independent of ρ. We thus have for all $\rho, \rho', \alpha, \; F(\alpha\rho)/F(\rho) = F(\alpha\rho')/F(\rho')$. But this is clearly equivalent to (16) of §64. Hence, by Theorem 4, $F(\rho) = k\rho^{\nu}$, and so

---

[41] See [10], p. 195; H. Schardin, *Comm. Pure Appl. Math.*, **7** (1954), pp. 223–43.

Theorem 9. For models of compressible flows to be dynamically similar at the same Mach number under all free stream conditions, the equation of state must have the special form

$$(36) \qquad\qquad p = k\rho^\gamma + \text{const.}$$

It is sufficient that γ be the same along all adiabatic curves. Generalizations to non-adiabatic flow are obvious.

Linearized Mach modeling. An interesting illustration of affine modeling is provided by the linearized (Prandtl-Glauert) approximation to steady compressible flow past thin bodies, already described in §§10–11. The perturbation ϕ in the velocity potential $U = ax + \phi(x, y, z)$ satisfies the differential equation (Ch. I, (14*)),

$$(37) \qquad\qquad (M^2 - 1)\phi_{xx} = \phi_{yy} + \phi_{zz}, \quad M = \frac{a}{c}.$$

In the *subsonic* case ($M < 1$), this is equivalent to $\nabla'^2 \phi = 0$, where $\nabla'^2 = \partial^2/\partial x'^2 + \partial^2/\partial y^2 + \partial^2/\partial z^2$, and $x' = x/\sqrt{1 - M^2}$, hence it reduces under an *affine* transformation to the *incompressible* case. In the *supersonic* case ($M > 1$), we have a similar reduction to

$$(37') \qquad\qquad \frac{\partial^2 \phi}{\partial x^2} = \frac{1}{k^2}\left\{\frac{\partial^2 \phi}{\partial y^2} + \frac{\partial^2 \phi}{\partial z^2}\right\}, \quad [k = \sqrt{M^2 - 1}],$$

which is the *wave* equation. In either case we get *affinely similar flow at corresponding Mach numbers around all affinely equivalent models.* The sonic case must be considered independently ([10], Sec. 9.6).

Thus, varying M is equivalent (at least in theory)[42] to changing the "thickness ratio" for affinely similar flows. Hence, besides the ordinary Mach modeling of (32), one can scale two perpendicular directions independently, much as in the theory of "nearly horizontal" flows.

Binary collision modeling.[42a] The transformation of distance and density in reciprocal ratio, with conservation of velocity and temperature,

$$x_i \to \alpha x_i, \quad t \to \alpha t, \quad u_i \to u_i, \quad \rho \to \rho/\alpha, \quad \theta \to \theta,$$

has some extraordinary properties. In a calorically perfect gas (§§3, 14), it conserves the specific heat C_v, the adiabatic constant γ, and the ambient sound speed C. Hence it conserves the Mach number $M = V/C$.

It also conforms to the kinetic theory of gases, insofar as only binary molecular collisions need be considered. Hence it conserves the viscosity μ, the conductivity κ, and changes the molecular mean free path λ in the ratio $1 : \alpha$. Hence it also conserves the Reynolds number $R = VL\rho/\mu$, the

---

[42] In view of §12 the theory may well be viewed with some caution. It involves no shock waves.

[42a] Unpublished work by the author and J. Eckerman of AVCO Corp.

Prandtl number $Pr = C_p \mu / \kappa$, and the Knudsen number λ / L. Thus, it models compressibility and shock wave effects, viscous effects, the temperature rise due to boundary layer heating, and rarefied gas (large mean free path) effects.

Finally, it preserves all second-order reactions of chemical kinetics; hence it models many of the phenomena, discussed in §34, which fall outside the scope of continuum mechanics. Otherwise, it has the great advantage of enabling one to reproduce to scale many aerothermodynamic effects in the upper atmosphere, by small scale model tests performed near the earth's surface.

74. Asymptotic scaling

The possibility of affine modeling—as in slender body theory—can be formally subsumed under dimensional analysis, by the device of attributing different "dimensions" to lengths in different directions.[43] However, a much more rational treatment is provided by inspectional analysis, which usually reveals such "dimensional analysis" as tantamount to singular *perturbation theory*—i.e. to asymptotic inspectional analysis.

We have discussed the case of linearized Mach modeling. We shall now give some illustrations of the same idea.

Perhaps the most important illustration is furnished by Prandtl's *boundary-layer* equations for laminar flow near a smooth solid boundary (§27). Thus, steady plane boundary-layer flow is determined (Ch. II, (14)) by

$$(38) \qquad u \frac{\partial u}{\partial x} + v \frac{\partial u}{\partial y} + \frac{1}{\rho} \frac{\partial p}{\partial x} = \nu \frac{\partial^2 u}{\partial y^2}, \qquad \frac{\partial u}{\partial x} + \frac{\partial v}{\partial y} = 0,$$

and the boundary conditions $u(x, 0) = 0$, $u(x, \infty) = u_\infty$. These formulas, derived under the approximation of vanishing boundary-layer thickness, are invariant under the group of affine transformations of the form

$$(39) \qquad x \to \beta^2 x, \qquad y \to \beta y, \qquad u \to u, \qquad v \to \beta^{-1} v,$$

as well as under the group defining Reynolds scaling.

Another example is furnished by the theory of irrotational gravity waves in shallow basins of gradually varying depth h. To a lowest order approximation, the mean particle velocity $u(x, t)$ of these "shallow water waves" [57, Sec. 2.2] satisfies

$$(40) \qquad u_{tt} = g(hu)_{xx}$$

---

[43] See W. Williams, *Phil. Mag.*, *34* (1892), 234–71; P. Moon and D. E. Spencer, *J. Franklin Inst.*, *248* (1949), 495–522. Often perturbation methods themselves cannot be justified rigorously.

in two-dimensional motion. A partial inspectional analysis shows that (40) is invariant under

(41a) $$t \to \beta t, \qquad h \to \beta^{-2} h,$$

for any $\beta > 0$. Since (40) is homogeneous and linear, (40) is also invariant under

(41b) $$u \to \delta u + \epsilon, \quad \text{for any } \delta > 0 \text{ and } \epsilon.$$

As already observed in §15, shallow water waves are represented to a next approximation by the equations for polytropic flow with $\gamma = 2$. It follows by §73 above that they are invariant under all changes of scale of the form

(41c) $$x \to \alpha x, \quad t \to \alpha t, \quad u \to u, \quad \rho \to k\rho.$$

In Chapter V we shall show how to exploit such groups to obtain explicit special solutions of boundary-value problems.

For the present we are primarily concerned with applications to the theory of modeling. Here an important application is to justifying distortion of the vertical scale[44] in hydraulic models. We shall discuss this practice in §77.

75. Wind-tunnels

Today many major laboratories have as their primary responsibility the carrying out and interpretation of model experiments. The practices followed have evolved out of simple ideas about similarity, like those described in this chapter. However, these original simple ideas have become profoundly modified through this evolution, and a discussion of modeling which did not at least indicate some practical aspects of the subject would be very misleading. Therefore, we shall conclude by giving a brief historical survey of the development of techniques for specific kinds of model experiments.

Wind-tunnels are probably the most highly developed facilities for model experiments. Originally, wind-tunnel data were necessarily interpreted through simply inertial scaling, as measurements of C_D, C_L, C_M for a given obstacle or wing *shape* at a given angle of attack. Moreover, the raw data required "tare corrections" for supports, as well as corrections for wall effects and pressure drop (§102). Since tests were made at various Reynolds numbers, discrepancies often arose, especially in the neighborhood of R_{crit}, which were sometimes suspected by rival laboratories to be experimental errors.[45]

---

[44] A. Craya [47]; A. T. Doodson, [51, p. 148].

[45] Personal recollection of Prof. W. Prager. Model roughness will, of course, also be a relevant parameter.

To obtain equal R with small models, an ingenious idea consists in using *variable density* wind-tunnels ([4], Sec. 102); since μ is pressure-independent, $R = VL/\nu = \rho VL/\mu$ is proportional to the density. However, experience shows that the ideas of §71 are also fallible in the sense that wind-tunnel *turbulence* may affect measured $C_D(R)$. To account for this, the idea has been suggested that to any given wind-tunnel one should assign a factor λ such that the "effective" Reynolds number R_{eff} is $\lambda VL/\nu$ in that tunnel. Today with the development of low-turbulence and full-scale wind-tunnels, not to mention free-flight testing and standardized profile families, low-velocity wind-tunnel testing has lost much of its uncertainty.

The technique of using *supersonic* wind-tunnels is much more recent. The first such tunnels were operated during World War II, and their operation was plagued by the occurrence of unforeseen "condensation shocks" and even snow—which can hardly be treated by simple dimensional analysis.

Dimensional analysis suggests that in supersonic wind-tunnels (and in exterior ballistics) one should plot $C_D(M, R)$. However, in practice the Reynolds number seems to be of secondary importance, though the contrary view has been widely disseminated:[46] Thus, skin friction is commonly only about 10% of the total resistance of a bullet; this is the component most likely to be influenced by viscosity and surface roughness. However, R_{eff} does also influence various minor phenomena, such as the thickness of shock waves and the λ-shocks discovered by Ackeret.[47]

76. Ship towing tanks

The most important application of Froude scaling is to model tests in ship towing tanks—though it is also used in model studies of water waves and seiches, in studies of water entry (§78), and is involved in hydraulic turbines having a free surface.[48]

There is a long-standing controversy as to whether Reech or Froude should get the credit for the idea of "Froude modeling" in ship resistance model tests. As the facts are rather curious, I shall state them.

Reech[49] certainly proposed, in 1831, exactly what is commonly referred to as "Froude's law," namely, testing ship models at equal Froude number, and estimating full-scale resistance by the similarity transformation

---

[46] The theoretical conclusions in *The Mechanical Properties of Fluids*, Blackie and Co., 1935, are based on a single table on p. 37 of [3]. Nothing remotely resembling the data of this table has been found in ballistic experiments performed in the United States.

[47] J. Ackeret, *Mitt. Inst. Aerodynamik Zurich*, No. 10, 1944.

[48] See J. Bertrand, *J. de l'Ec. Polyt.*, 19 (1948), 189–97.

[49] F. Reech, *Cours de l'Ecole d'Application du Génie Maritime*, Lorient, 1831.

(22). Froude's great merit consists in having gone beyond this simple law. In most merchant ships ninety per cent of the resistance is frictional, and hence susceptible to Reynolds and not to Froude modeling. In order to estimate hull resistance from model tests, one must separate ship resistance into its frictional and wave components. This was first proposed in 1874 by Froude;[50] the basic assumption can be interpreted as the formula

$$(42) \qquad C_D = C_W(F) + C_f(R).$$

For merchant ships $C_f(R)$ is usually predominant!

However, the exact model "laws" are still far from clear. Since 1935, $C_f(R)$ itself has usually been analyzed into "skin friction" and "form drag" (wake or eddy resistance).

Extrapolation of form drag from model to full-scale is especially subject to personal judgment. Since about 1945, it has become common practice to roughen models artificially, so as to obtain an "effective Reynolds number" R_{eff} and form drag coefficient more nearly resembling that of an actual ship. There is no scientific principle, known to the author, for determining just how much roughness is required—especially as "fouling" greatly alters the frictional resistance of a real ship during its life.

Recently it has been suggested that ship wave resistance should be *calculated* theoretically, and the friction (and form) drag regarded as residual. Such calculations have not yet been made: the nonlinear "free boundary" condition makes them formidable.

In the linearized "thin ship" approximation (§74), the theoretical $C_w(F)$ reduces to a quintuple integral (the Michell integral). This has been calculated in a few simple cases. However, it is not yet clear whether the difference between the calculated $C_w(F)$ and the observed $C_D - C_f(R)$ is due to nonlinearity or to the wake.[51]

77. River and harbor models

Still more dependent on practical experience and inspired intuition is the interpretation of model data involving harbors, rivers, estuaries, dams, spillways, and so on.[52] The modeling of fluid flow in so-called "fixed beds" is complicated enough; that of scouring and sedimentation in "movable bed" models is even further removed from the simple mathematical concept of inspectional analysis.

[50] *Trans. Inst. Nav. Arch.*, *15* (1874), 36–59.

[51] G. Birkhoff, J. Kotik, and B. V. Korvin-Kroukovsky, *Trans. Soc. Nav. Arch. Marine Eng.*, *62* (1954), 359–96.

[52] For an authoritative conventional discussion see H. Warnock in Ch. II of *Engineering Hydraulics*, Hunter Rouse ed. See also *Am. Soc. Civ. Eng. Manual of Engineering Practice*, No. 25.

For studies of fluid flow in scaled-down fixed beds, Froude modeling can be used to a first approximation. That is to say, if the reduction in scale is $L:1$, the reduction in velocities will be $\sqrt{L}:1$, and that in volumetric discharge $L^{5/2}:1$, as suggested in §72, all very approximately. (That in tidal periods will also be $\sqrt{L}:1$.)

However, experience with Froude models quickly leads one to recognize various limitations. Thus, wave attenuation and other viscous effects are exaggerated in small models. Waves in small harbor models do not break like full-scale waves: capillary effects[53] exert a decisive influence. Moreover, the entrainment of air in small models of spillways and waterfalls is much less than in full-scale reality.[54]

Most important is the fact that (§71) in models, viscous forces are relatively greater, and hence turbulence ("eddy viscosity") is less. The usual cure is to *induce* turbulence in models by artificially roughening surfaces or even by obstructing fluid motion with upright plates or coarse wire meshes. This increases the eddy viscosity so that the viscous forces in the model are even greater, relatively, than they would be otherwise.

However, at sufficiently high Reynolds numbers this somewhat paradoxical procedure may be partially justified by inspectional analysis, as was pointed out to the author by Dr. S. K. Roy. The viscous forces are then much less than the "Reynolds stresses" $\overline{u_i' u_j'}$, where u' denotes the turbulent vector velocity, and a bar denotes averaging (see [4], p. 192). Hence, if the relative turbulence is the same at all points, the mean velocity distribution may be expected to be similar.

The transport of solid particles (dirt, sand, gravel) by moving water is frequently studied in "movable bed" models because of its great practical importance for rivers, harbors, and estuaries. The use of such models is a highly individualized art which involves very complicated considerations.[55] Froude modeling roughly scales the velocities due to gravity head and (in harbor models) wave motions as well, but only in the turbulent regime or if viscosity is negligible.[56]

The relative particle size is frequently exaggerated in the model, partly to avoid cohesive forces, partly to preserve R, and partly for ease

[53] To model these, one must preserve the "Weber number" $W = \gamma/\rho V^2 L$, where γ is the surface tension (see Bashforth and Adams, *Capillary Action*, (1883)).

[54] See L. Escande, *Génie Civile*, 16 Dec. 1939; C. Camichel and L. Escande, *Similitude Hydrodynamique et Technique des Modèles Reduits*, Paris, 1938 (scarce).

[55] A lengthy discussion of some complications may be found in *Proc. Am. Soc. Civ. Eng.*, 71 (1944), *Trans. No. 3*, Part 2. In certain tidal models the Coriolis force must be modeled by changing river curvatures; see [44], p. 78.

[56] One can exaggerate the tilt of the bottom of the river or estuary to obtain appropriate mean velocities of flow when R is small. Roughness is an important factor.

of fabrication. This increased size tends to keep the model particles from being stirred up by the water—a tendency which is compensated for[57] by reducing their negative buoyancy $\rho_1 - \rho$.

The common use of *different horizontal and vertical scales* in such models also deserves comment. In England it is usual to exaggerate the vertical scale (following Reynolds and Gibson) to avoid the tendency to excessive shallowness. The validity of this exaggeration is often disputed[58] in France, where there is tendency to use larger models instead. This can be interpreted as a form of asymptotic scaling (§74).

In practice, theoretical considerations are seldom invoked in hydraulic model studies of rivers and harbors. Reliance is placed on reproducing various aspects of the observed behavior under actual conditions. It is hoped that variations in behavior due to altered conditions will then also be reproduced to scale—even though there is no rational argument to support this hope.

78. Modeling of water entry

For underwater ballistics it may be important to model the phenomena of "surface seal" and "down-nosing" which accompany entry into water, as stated in §53. This raises the problem of trying to reproduce these effects to scale.

On the basis of experimental analogy one might be tempted simply to use reduced pressure Froude modeling, preserving F and Q or Q^*; this has indeed been proposed. However, I am glad to state that, for once, the correct solution as regards surface seal appears to have been given, not by an engineer on the basis of physical experience, but by a "mathematician in his armchair," using inspectional analysis—namely, myself.[59] The answer was suggested by the following considerations.

A little reflection makes it plausible that it is the *air density* which causes surface seal—there being a reduction of $\rho'v'^2/2$ ($\rho' =$ air density, $v' =$ air velocity) in the pressure at the neck of the cavity, which causes the splash and neck to contract. This is not scaled if ρ' is reduced by lowering pressure; while if one does not lower the pressure, presumably the size of the bubble after seal is not modeled.

[57] See C. Camichel and L. Escande, *Comptes Rendus, 199* (1934), 992. An excellent review of scaling laws for silt transport is given by H. A. Einstein and R. Muller, *Schweiz. Archiv, 5* (1939), No. 8.

[58] C. Camichel, E. Fischer, and L. Escande, *Comptes Rendus, 199* (1934), 594. The experimental counterexample given by them is not one of nearly horizontal flow.

[59] G. Birkhoff, *Modeling of Entry into Water*, Applied Math. Panel, National Defense Research Council, May, 1945 (declassified). John C. Waugh and G. W. Stubstad, *Water-Entry Cavity Modeling*, Navord Rep. 5365, Dec., 1957. For modeling of entry of solids into water, see also A. May, *J. Appl. Phys., 19* (1948), 127–39; J. Levy, Rep. *E-12.19*, Hydrodynamics Lab., Caltech., Aug., 1956.

Since one is not dealing with water vapor, the equation of state is no longer (33), but is nearly

$$(43) \qquad\qquad p' = k\rho'^\gamma.$$

The same equation is satisfied (approximately) by underwater gas bubbles; unless we are dealing with saturated gas, we do not have $p' = p_v + k\rho'^\gamma$. For this reason we should use $Q*$ instead of Q.

Mathematically, the Equations of Continuity, State, and Motion are preserved under the transformation

$$x_1 = \alpha x, \qquad t_1 = \sqrt{\alpha}t, \qquad u_1 = \sqrt{\alpha}u, \qquad p_1 = \alpha p,$$
$$p_1' = \alpha p', \qquad \rho_1 = \rho, \qquad \rho_1' = \rho', \qquad k_1 = \alpha k,$$

where suffixes denote transformed variables. Hence, we could obtain a model if we could get a gas whose equation of state had the form $p_1' = \alpha k \rho_1'^{1.408}$, where $p' = k\rho'^{1.408}$ was that of air. For example, air at a low temperature would provide a good solution, but this is not practical from an engineering point of view.

It appears to be more practical to use Freon or some other "heavy gas" with $\gamma \neq 1.4$, but having a density several times that of air at reduced pressure and atmospheric temperature; this has been done on a small scale (cf. [31]). One can then only model ρ and $dp/d\rho$ in the gas.

The modeling procedure outlined above presumably does not model "down-nosing." If down-nosing is a viscous effect, as suggested in §53, it can only be modeled by preserving R, and this is impractical. However, since the maximum possible underpressure decreases to zero with $Q*$ (assuming tension $p < 0$ impossible for the durations involved), down-nosing should at least be mitigated by reduced pressure Froude modeling, with or without a heavy gas.

The need for a more careful analysis than that provided by ordinary dimensional analysis is also illustrated by the dimensionless parameter $N = \sqrt{F}\rho'/\rho$, which has recently[60] been shown to determine roughly whether entry into water is accompanied by surface seal or deep seal. The derivation of N as a criterion rests on the empirical fact that the time T_{ds} required for deep seal is roughly proportional to $\sqrt{L/g}$, combined with the fact (deducible from inspectional analysis of the inertial mechanism of surface seal) that the time T_{ss} of surface seal is proportional to $\rho L/\rho' V$, where L is a typical length and V a typical velocity. The condition $T_{ds} > T_{ss}$ for surface seal therefore assumes the form $N > N_{\text{crit}}$.

Using ordinarily dimensional analysis, one would have argued that the average pressure differences causing surface seal and deep seal, respectively, should be proportional to $\rho' V^2/2$ and $2\rho g L$, suggesting the use of the dimensionless ratio $N' = F\rho'/\rho$ as the criterion for surface seal. This disagrees strikingly with observation.

[60] G. Birkhoff and R. Isaacs, *Transient Cavities in Air-Water Entry*, Navord Rep. 1490, Jan. 1951.

V. Groups and Fluid Mechanics

79. Introduction

In Chapter IV the group concept was shown to be valuable in fluid mechanics in three ways. First, through "inspectional analysis" it provides a mathematical basis for the use of models, which is far more adequate than the "dimensional analysis" commonly appealed to. Second, it provides tests for the validity for mathematical theories of hydrodynamics, even in cases where it is impossible to integrate the partial differential equations involved in such theories. And finally, like dimensional analysis (but more generally), it often permits one to reduce the number of parameters which need to be considered; thus it provides important simplifications.

I shall now discuss its utility as a method for *integrating* the differential equations of fluid mechanics and, indeed, of mathematical physics generally. Most of what I shall have to say in this connection has been said in one way or another somewhere before in the literature. But if, as I believe, we have only begun to explore the applications of the group concept to differential equations, it seems worth collecting these ideas in one place.

I shall first describe what may be called the method of *search for symmetric solutions* of partial differential equations. Suppose that a system Σ of partial differential equations is invariant under a group G of the independent *and* dependent variables involved. The method consists in searching for a solution which is invariant under a subgroup of G. In other words, it consists in looking for *self-similar* solutions possessing *interal symmetry* with respect to G.

This method has been applied so often to specific physical problems that it is surprising that it should not have been more explicitly recognized long ago.[1] I shall now illustrate its effectiveness by various specific applications.

80. Symmetric solutions of heat equation

The method of "search for symmetric solutions" is applicable to continuum physics generally. A simple application is to the diffusion

---

[1] It was first formulated by K. Bechert [58]. A more explicit formulation has been given by L. I. Sedov, [69] and [56], Ch. IV, §1. See also K. P. Staniukovitch [70] and [71]. Russian writers refer to "automodel" solutions.

equation; this will be treated first. The Navier-Stokes equations reduce to the diffusion equation in parallel flow,[2] but the heat conduction application is best known. Because heat conduction and viscous momentum transfer have the same symmetry group, one can apply analogous considerations to some problems involving heat conduction *and* convection. For instance, one can treat problems involving change of phase at moving boundaries (Stefan problem), such as the growth of spherical steam bubbles in uniformly superheated water.

Accordingly, we consider the diffusion of heat from a *point*, in a medium of constant thermal diffusivity κ. The equation of heat conduction in solids is

$$(1) \qquad \frac{\partial U}{\partial t} = \kappa \nabla^2 U = \kappa \sum_{i=1}^{n} \frac{\partial^2 U}{\partial x_i^2}, \quad (n = 1, 2, \text{ or } 3),$$

for the temperature U at the point $x = (x_1, \cdots, x_n)$ at time t.

To preserve rotational symmetry, we look for a solution of the form $U(r, t)$ where $r^2 = \sum_{i=1}^{n} x_i^2$. This exhausts the purely *geometrical symmetry* of the problem. But we also have *physical symmetry*, in the sense that the differential equation (1) is invariant under the group of transformations of space-time and temperature

$$(1^*) \qquad r' = \alpha r, \qquad t' = \alpha^2 t, \qquad U' = \beta U + \gamma,$$

depending on the three arbitrary parameters α, β, γ. This group extends the classical Law of Times, according to which the time required for thermal effects to propagate is proportional to the square of the distances involved. For any positive number m, the three-parameter group (1^*) contains a one-parameter subgroup, defined by

$$(2) \qquad r' = \alpha r, \qquad t' = \alpha^2 t, \qquad U' = \alpha^m U.$$

Since $\gamma = 0$, this subgroup preserves the boundary condition $U(\infty, t) = 0$. *We look for solutions $U(r, t)$ of (1) which are invariant under subgroups (2).*

In the present case, since the group (2) consists of scalar multiplications, we can use the Pi Theorem. The variables $\chi = r^2/t$ and $U/t^{m/2}$ are invariant under (2). Hence by the Pi Theorem any solution of (1) which is invariant under (2) must be of the form

$$(3) \qquad U = t^{m/2} f(\chi), \qquad \chi = r^2/t.$$

We will show in §89 that (1) always does have solutions of the symmetric ("automodel" or "self-similar") form (3), at least locally.

[2] [7], Secs. 345–7. The results of this section were presented in [59], before I knew of the papers cited in footnote 1. For the application to steam bubble growth, see G. Birkhoff, W. A. Horning, and R. Margulies, *Physics of Fluids, 1* (1958), 201–4.

For the present we content ourselves with working out the special case under consideration. Substituting from (3) into (1), and dividing out a suitable power of t, we get

$$\text{(4)} \qquad 4\kappa\chi f'' + (2\kappa n + \chi)f' - \frac{m}{2} f = 0.$$

Use of the variable $\xi = r^2/4\kappa t$ (which is *dimensionless* in the usual sense of being invariant under the group (22) of Ch. IV) leads to the simpler relations

$$\text{(4')} \qquad U = t^{m/2} F(\xi), \quad \text{where} \quad \xi F_{\xi\xi} + \left(\xi + \frac{n}{2}\right)F_\xi - \frac{m}{2} F = 0.$$

This is equivalent to the confluent hypergeometric equation,[3] under the substitution $x = -\xi$. This is however not the main point, which is that *solutions of (1) can be found by integrating an ordinary differential equation*, which can always be done numerically.

Not all "symmetric" solutions of (1) (i.e. of (4)) have equal physical interest. Interest is largely confined to solutions for which $U \to 0$ as $r \to \infty$, so that $\lim_{\chi \to \infty} f(\chi) = 0$. We are also interested in the *total heat*, which is proportional to

$$\int_0^\infty U(r, t) r^{n-1}\, dr, \quad \text{and hence to} \quad \alpha^m \alpha^n = \alpha^{m+n}, \quad \text{or to} \quad t^{(m+n)/2}.$$

One special case of interest is that of *constant* total heat, corresponding to diffusion of a fixed quantity of thermal energy, initially at the origin. Here $m = -n$; if we set $n = 2h$, then (4') reduces, after collecting terms, to

$$\text{(4*)} \qquad 0 = \xi F_{\xi\xi} + (\xi + h)F_\xi + hF = \xi(F_\xi + F)_\xi + h(F_\xi + F), \quad h = n/2.$$

Equation (4*) can be integrated in closed form. To make $U(\infty, t) = 0$, F and F_ξ must tend to zero as $\xi \to +\infty$, and so $F(\xi) = e^{-\xi}$. This gives Laplace's solution:

$$\text{(5)} \qquad U = Ct^{-n/2} e^{-r^2/4\kappa t},$$

derived in most textbooks by using Fourier transforms.

Another interesting case is a *point source* producing heat (by chemical or radioactive process) at a constant rate, beginning at $t = 0$. Here $m + n = 2$, or $m = 2 - n$, whence (4') becomes

$$\xi F_{\xi\xi} + (\xi + h)F_\xi + (1 - h)F = 0, \quad h = \frac{n}{2} = \frac{1-m}{2}.$$

The integrals of this form of the confluent hypergeometric equation

[3] See E. Kamke, *Differentialgleichungen*, 3d ed. Leipzig, 1944, p. 426, Formula 2.107.

(obtained by other means) can also be expressed in a closed form.[4] However this is again not as important as the fact that the differential equation is ordinary.

81. Spiral viscous flows

We next illustrate the method of "search for symmetric solutions" by the classical case of the "spiral flows" of incompressible viscous fluids. The final formulas were first obtained by Jeffery[5] and Hamel. The degenerate special cases of radial flows in channels and of circular Couette flows have the greatest importance for applications. However, we shall treat the general case because of its mathematical interest.

It is well known that in incompressible plane flows the equation of continuity is equivalent to the introduction of a "stream function" $V = \int (u\,dy - v\,dx)$, such that $(\partial V/\partial y,\ -\partial V/\partial x)$ is the vector velocity. Then $-\nabla^2 V = \partial v/\partial x - \partial u/\partial y$ is the *vorticity*. Moreover, the Navier-Stokes equations of motion are equivalent[6] for such plane flows to

$$(6) \qquad \nu\nabla^4 V = \frac{\partial(\nabla^2 V,\ V)}{\partial(x,\ y)} = \frac{1}{r}\frac{\partial(\nabla^2 V,\ V)}{\partial(r,\ \theta)}$$

$$= \frac{1}{r}\left[\frac{\partial V}{\partial\theta}\frac{\partial(\nabla^2 V)}{\partial r} - \frac{\partial V}{\partial r}\frac{\partial(\nabla^2 V)}{\partial\theta}\right].$$

Here $\partial(p,\ q)/\partial(x,\ y) = p_x q_y - q_x p_y$ is the ordinary notation for the Jacobian; ν is, as usual, the kinematic viscosity μ/ρ.

Dimensional analysis tells us that under geometrically similar conditions the behavior of incompressible viscous fluids will be a function of the dimensionless parameter R alone. We now look for *self-similar* plane flows, under one-parameter subgroups of the similarity group

$$r' = e^\alpha r, \qquad \theta' = \theta + \beta.$$

That is, we consider flows invariant under some *spiral* subgroup

$$(7) \qquad\qquad r' = e^\alpha r, \qquad \theta' = \theta + c\alpha,$$

where the parameter c characterizes the spiral.

The transformations (7) are self-similitudes of the plane. Since ρ is fixed, they will make the motion self-similar at constant R if and only if ru_r and ru_θ are the same at corresponding points. But *differences* in the

---

[4] See Carslaw and Jaeger, *Conduction of Heat in Solids*, Art. 104. The case $n = 2$ is especially obvious, as then $(1 - h) = 0$.

[5] G. B. Jeffery, *Proc. Lond. Math. Soc.*, *14* (1915), 327–38; G. Hamel, *Jahr. Deutsche Math. Ver.*, *25* (1916), 34–60.

[6] This result may be derived from Chapter II, (11), since curl $\mathbf{g} = 0$ in a conservative field, and since $\boldsymbol{\xi}\cdot\nabla = 0$ because $\xi_1 = \xi_2 = \partial/\partial z = 0$. The result may also be found in [8], p. 573, Ex. 7.

stream function V are proportional to distances times velocity, since $dV = (\partial V/\partial x)dx + (\partial V/\partial y)dy$. Therefore, differences in V will be *invariant* under the spiral group (7). Hence, given $(r, \theta) = (e^\lambda, \theta)$ in polar coordinates, by transforming under (7) with $\alpha = -\lambda$, we get

$$(8) \quad V(e^\lambda, \theta) - V(e^\lambda, c\lambda) = V(1, \theta - c\lambda) - V(1, 0) = F(\chi), \quad \chi = \theta - c\lambda.$$

For flows self-similar under (7), for the same reason, equal changes in λ will produce equal changes in $V(e^{-\lambda}; c\lambda)$; hence $V(e^\lambda, c\lambda) = a\lambda + b$ is a linear function of λ. (The constant b does not affect the velocity, and may be set equal to zero.) Combining this result with (8), we get

$$(9) \quad V(r, \theta) = a\lambda + F(\chi), \quad \lambda = \ln r, \quad \chi = \theta - c\lambda.$$

This expresses the flow in terms of an arbitrary constant a and a function F of one variable. The most interesting case is that of *spiral streamlines*, when $a = 0$ in (9).

As before, we now substitute in the general differential equation (6); the calculations follow. In general, we have

$$(10) \quad \nabla^2 = \frac{1}{r^2}\left(\frac{\partial^2}{\partial\lambda^2} + \frac{\partial^2}{\partial\theta^2}\right), \quad \text{whence} \quad \nabla^2 V = e^{-2\lambda}(c^2 + 1)F'' \quad \text{by (9)}.$$

Differentiating again, we get

$$(10') \quad \frac{\partial}{\partial\lambda}(\nabla^2 V) = -e^{-2\lambda}(c^2 + 1)[cF''' + 2F''], \qquad \frac{\partial}{\partial\theta}(\nabla^2 V) = e^{-2\lambda}(c^2 + 1)F'''$$

The area-ratio formula $\partial/\partial(x, y) = r^{-2}\partial/\partial(\lambda, \theta)$ for Jacobians shows that the Navier-Stokes equations (6) are equivalent to

$$\frac{\nu}{r^2}\left[\left(\frac{\partial^2}{\partial\lambda^2} + \frac{\partial^2}{\partial\theta^2}\right)\nabla^2 V\right] = \frac{1}{r^2}\left[\frac{\partial V}{\partial\theta}\frac{\partial(\nabla^2 V)}{\partial\lambda} - \frac{\partial V}{\partial\lambda}\frac{\partial(\nabla^2 V)}{\partial\theta}\right].$$

Carrying out the operations indicated, using (10)–(10'), and taking out the common factor $(c^2 + 1)/r^4 = e^{-4\lambda}(c^2 + 1)$, we get

$$(11) \quad \nu[(c^2 + 1)F^{IV} + 4cF''' + 4F''] + aF''' + 2F'F'' = 0.$$

This is the ordinary differential equation obtained by Oseen;[7] it would be difficult to find another equally simple motivation. Equation (11) can be made somewhat less repulsive through the substitution $G = F'$; moreover, it is satisfied whenever $F'' = 0$. In any case, solutions can be found by numerical integration.

---

[7] See C. W. Oseen, *Arkiv for Mat.*, *I–II*, 1927–8, or [68], Ch. II. For asymptotic behavior with small ν, see G. Kuerti, *J. Math. Phys. MIT*, *30* (1951), 106–15.

82. Boundary layers on wedges

We next consider the problem of integrating the laminary boundary-layer equations for steady plane flow, already introduced in §27. These equations are

$$(12) \qquad u \frac{\partial u}{\partial x} + v \frac{\partial u}{\partial y} = u_\infty \frac{du_\infty}{dx} + \nu \frac{\partial^2 u}{\partial y^2}, \quad \frac{\partial u}{\partial x} + \frac{\partial v}{\partial y} = 0,$$

subject to the boundary conditions

$$(13) \qquad u = v = 0 \quad \text{on} \quad y = 0, \quad x \geqq 0$$

and

$$(13') \qquad \lim_{y \to +\infty} u(x, y) = u_\infty(x).$$

As observed in §74, the above equations were derived under *asymptotic* approximations. This suggests that we consider x and y as independent dimensions, and look for symmetries under non-trivial subgroups of the four-parameter group of affine transformations

$$(14) \qquad x \to \alpha x, \quad y \to \beta y, \quad u \to \gamma u, \quad v \to \delta v.$$

A hopeful case to consider is that of flow past an infinite symmetric *wedge*. In this case an elementary conformal transformation reveals the Euler flow outside the boundary layer to satisfy[8] $u_\infty(x) = c x^m$ for suitable constants c and m. The case $m=0$ refers to a flat plate parallel to the stream; the case $m=1/2$ refers to flow impinging on a flat plate.

Testing (12) and (13') for invariance under (14), with $u_\infty(x) = c x^m$, we are led to the one-parameter subgroup defined by

$$(14^*) \qquad \beta = \alpha^{(1-m)/2}, \quad \gamma = \alpha^m \text{ (trivially)}, \quad \delta = \alpha^{m-1}\beta = 1/\beta.$$

The variable $\eta = (u_\infty/x)^{1/2}y$ is invariant under this subgroup; since it carries $V = \int u \, dy$ into $\alpha^{(m+1)/2}V$, the function f defined by $V = f(x, y)$ is also invariant. Hence, we look for a solution of the special form $V = x^{(m+1)/2}f(\eta)$—i.e. for a solution invariant under the subgroup (14^*).

Any V of this form will satisfy (13)–(13') and the second equation of (12) if $f'(\infty) = c$. To satisfy the remaining equation, it is necessary and sufficient that $f(\eta)$ satisfy

$$(15) \qquad m(f'^2 - 1) - \frac{m+1}{2}ff'' = \nu f'''.$$

This can be integrated numerically,[9] relative to the boundary conditions $f(0) = f'(0) = 0, f'(\infty) = c$.

---

[8] The argument is due to Falkner and Skan [64]; see also [4], Art. 64. The case of a flat plate was first treated by Blasius [61]. See also Weyl [74]; Geis [65].

[9] See D. R. Hartree, *Proc. Camb. Phil. Soc.*, *33* (1937), 223–9; S. Goldstein, ibid., *35* (1939), 338–41; K. Stewartson, ibid., *50* (1954), 454–65.

83. Viscous jets and wakes

Using considerations like those above, one can calculate the asymptotic velocity profiles of laminar viscous jets, in the boundary-layer approximation, for both plane and axially symmetric flow. Because of the affine invariance of the boundary-layer and "continuity" equations (12), we look for velocity profiles satisfying the Similarity Hypothesis

$$(16) \qquad u = x^{-p} f(\eta), \quad \eta = \frac{y}{x^q},$$

where y denotes distance from the x-axis, in plane or space. For equations (12) to be invariant under (16), it is necessary and sufficient that $2q = p+1$. We omit the calculation,[10] but note that it confirms the result $\beta = \alpha^{(1-m)/2} = \alpha^q$ of (14*), in the case $p = -m$ treated in §82.

To determine p, one must also utilize a law of conservation of total *jet momentum*, quite like the law of conservation of wake momentum treated in §57. This conservation law is equivalent to $2p = q$ in the plane, and $p = q$ in space, if (16) is assumed.

Solving the preceding equations, we obtain $p = q = 1$ in space. This is remarkable because the full Navier-Stokes equations are invariant under the special group of similarities obtained, as was first observed by Yatseev and Squire.[11] The Navier-Stokes equations are equivalent, in spherical coordinates, to

$$(17) \qquad f^2 = 4\gamma f + 2(1-\gamma^2)f' - 2(c_1\gamma^2 + c_2\gamma + c_3), \quad \gamma = \frac{x}{r} = \cos\theta,$$

where c_1, c_2, c_3 are constants of integration. Moreover, the natural physical boundary conditions imply $c_1 = c_2 = c_3 = 0$; in this case, one can easily integrate (17), getting

$$(17') \qquad f = \frac{2\sin^2\theta}{a+1-\cos\theta},$$

for arbitrary a. The behavior of these solutions "in the large" will be discussed in §89.

Laminar viscous wakes can be treated analogously if one lets u be the perturbation in the free-stream velocity U, so that $U + u$ is the local velocity. In this case one must combine the Similarity Hypothesis (16) with *wake momentum* conservation (§57), getting $p = q = 1/2$ in plane wakes, and $p = 1$, $q = 1/2$ in space. The velocity profiles can also be calculated, always in the laminar boundary-layer approximation.

Turbulent jets and wakes have been treated in much the same way.

[10] [15], p. 271.

[11] V. L. Yatseev, *Zh. Eksp. Teor. Fiz.*, 20 (1950), 1031–4; H. B. Squire, *QJMAM*, 4 (1951), 321–9. See [15], p. 278, for the boundary conditions.

However, it is now generally conceded that the "mixing length" assumptions on turbulence similarity, made in published theories, are very dubious (see [15], Ch. XIV, §11).

84. Prandtl-Meyer expansion wedges

To illustrate further the applicability of the method of "search for symmetric solutions" to problems of continuum physics, we now turn our attention to permanent irrotational flows of compressible nonviscous fluids. The differential equations of such flows are, as we have seen in §73, invariant under the one-parameter group of Mach modeling

$$(18) \qquad x_i \to \alpha x_i, \qquad t \to \alpha t, \qquad p, \rho, u \text{ unchanged.}$$

The method of search for symmetric solutions suggests looking for flows which are invariant under (18); stationary flows will be invariant also under the group

$$(18') \qquad\qquad t \to t + \tau;$$

all other variables are unchanged.

In treating flows invariant under (18)–(18'), it is convenient to use polar coordinates (r, θ) and spherical coordinates (r, θ, ϕ) respectively, and to let u_r and u_θ be the associated radial and angular velocity components. We consider only the case $u_\phi = 0$ of no swirl, in steady (irrotational) plane and axially symmetric flow.

For such flows, the assumption of invariance under (18)–(18') means that the following quantities are functions of the angular (colatitude) variable alone:

$$u_r = g(\theta), \qquad u_\theta = h(\theta), \qquad \rho = \rho(\theta),$$

$$(19)$$

$$p = f(\rho), \qquad \frac{dp}{d\rho} = f'(\rho) = c^2.$$

Plane flows satisfying (19) are called *wedge* flows; they will be generalized in §92 (see Fig. 26 there). Space flows which satisfy (19) are called axially symmetric *conical* flows.

The assumption of irrotationality amounts to $\oint u_r \, dr + u_r r d\theta = 0$ for all closed curves, whence $0 = \partial(r u_\theta)/\partial r - \partial u_r/\partial \theta = h - g'$, and we get

$$(20) \qquad\qquad h = g'.$$

Since the flow is irrotational, the Equations of Motion are equivalent to the Bernoulli equation, which we can write as

$$(21) \qquad \tfrac{1}{2}u^2 + \int \frac{dp}{\rho} = \tfrac{1}{2}(g^2 + g'^2) + \int \frac{dp}{\rho} = \text{const.},$$

or, in differential resp. differentiated form:

$$(21') \qquad 0 = u\,du + \frac{dp}{\rho} = g'(g+g'') + c^2\left(\frac{\rho'}{\rho}\right).$$

With a *polytropic* equation of state $p = k\rho^\gamma + p_0$, $c^2 = \gamma k \rho^{\gamma-1}$, and so $\int dp/\rho = k\gamma\rho^{\gamma-1}/(\gamma-1) = c^2/(\gamma-1) + \text{const.}$; hence, in this case

$$(21^*) \qquad \tfrac{1}{2}(g^2 + g'^2) + \frac{c^2}{\gamma-1} = \text{const.} = C.$$

So far, everything is equally valid for conical flows.

With *wedge* flows, the Equation of Continuity, $\text{div}\,(\rho\boldsymbol{u}) = 0$, can be written in the form

$$0 = \frac{\partial}{\partial r}\,(\rho r u_r) + \frac{\partial}{\partial\theta}\,(\rho u_\theta) = \rho u_r + (\rho u_\theta)' = \rho(u_r + u_\theta') + \rho' u_\theta.$$

Substituting from (19)–(20), this is equivalent to

$$(22) \qquad (g+g'') + \left(\frac{\rho'}{\rho}\right) g' = 0.$$

Multiplying (22) by g' and subtracting from (21'), we get

$$(22') \qquad (c+g')(c-g')\left(\frac{\rho'}{\rho}\right) = 0.$$

Clearly (22)–(22') are equivalent to the equations of motion and continuity. We have two families of solutions.

Case I. $\rho' = 0$. Substituting in (22), $g'' + g = 0$, whence $u_r = A\,\cos\,(\theta-\alpha)$. Using (19)–(20) to get u_θ, we see that this is *uniform* flow with constant vector velocity.

Case II. $c^2 = g'^2$, or $\pm c = g'$. We see that the radii $\theta = \text{const.}$ are characteristics, in the sense that the component of velocity perpendicular to them is always the sound velocity c. This gives so-called *Prandtl-Meyer* [13] *expansion wedges*; they can fill wedge-shaped regions, with smooth transition at the boundaries into regions of uniform flow. Such regions are often seen in Schlieren photographs of actual flows; the hypothesis $\rho = \rho(\theta)$ thus gets direct experimental confirmation.

In the polytropic case, substituting $c^2 = g'^2$ in (21*), we get immediately the differential equation

$$(23) \qquad (\gamma+1)g'^2 + (\gamma-1)g^2 = 2(\gamma-1)C.$$

This can be easily integrated in closed form, the qualitative nature of the solutions in the adiabatic case $\gamma > 1$ being quite different from what it is if $\gamma = 1$, if $-1 < \gamma < 1$, or if $\gamma = -1$ (circular flow).

[13] [66]; T. Meyer, *VDI Forschungsheft, 62* (1908), 31–67.

In the general non-polytropic case, we still have (22) and $c = \pm g'$. By symmetry, it suffices to treat the case $c = g'$. Using (21) to solve for $\ln \rho = \psi(c) = \psi(g')$, we then get

$$(23') \qquad\qquad (g + g'') + g'g''\psi'(g') = 0,$$

which can be integrated by numerical quadrature ([4a], Sec. 7.1). (There is a singularity if $g'\psi'(g') = -1$.) This shows that *Prandtl-Meyer expansion wedges are mathematically possible with a general equation of state*.

85. Taylor-Maccoll conical flows

In n dimensions (the physically interesting case is of course $n = 3$) the Equation of Continuity for stationary axially symmetric flows assumes the form

$$0 = \frac{\partial}{\partial r} (\rho r^{n-1} \cos^{n-2} \theta \, u_r) + \frac{\partial}{\partial \theta} (\rho r^{n-2} \cos^{n-2} \theta \, u_\theta)$$

$$= \rho r^{n-2} (\cos^{n-2} \theta \, u_\theta)' + [(n-1)\rho u_r + \rho' u_\theta] r^{n-2} \cos^{n-2} \theta.$$

After dividing by $\rho r^{n-2} \cos^{n-2} \theta$, and substituting from (19)–(20) (we recall that equations (19)–(21′) are valid in space), we get

$$(24) \qquad (n-1)g - (n-2)g' \tan \theta + g'' + g' \left(\frac{\rho'}{\rho} \right) = 0$$

in place of (22). Here (21′) is still valid; it is equivalent to (21*) in the polytropic case. In this case, by (21)–(21*), we have

$$\left(\frac{\rho'}{\rho} \right) = -\frac{g'(g + g'')}{c^2} = \frac{g'(g + g'')}{(\gamma - 1)} \left[\frac{1}{2}(g^2 + g'^2) - C \right]^{-1}.$$

Substituting in (24), which can be written

$$(24') \qquad (g'' + g) + (n-2)(-g' \tan \theta + g) + g' \left(\frac{\rho'}{\rho} \right) = 0$$

we get

$$(25) \qquad (g'' + g) \left\{ 1 + \frac{g'^2}{(\gamma - 1)} \left[\frac{1}{2}(g^2 + g'^2) - C \right]^{-1} \right\} + (n-2)(-g' \tan \theta + g) = 0$$

This has been integrated numerically for $\gamma = 1.408$ (air) in a celebrated paper by G. I. Taylor and J. W. Maccoll [72].

In ballistic applications (see Fig. 24), one is interested in flows which are first uniform and parallel, then deflected by a conical shock wave of constant intensity according to the Rankine-Hugoniot equations (which are also invariant under (18)), followed by a region in which (25) holds,[14]

---

[14] In the plane analog, one has straight streamlines in this region (§84, Case I), as first observed by A. Busemann, *ZaMM*, 9 (1929), 496–8.

until the flow becomes purely radial ($g' = 0$)—i.e., tangent to an idealized projectile cone head, making a vertex semi-angle β with the flow direction. For given β, γ one can integrate (25) numerically as an initial value problem for $g'(\beta) = 0$ and each dimensionless ratio $g^2(\beta)/C$ in (21*). There will be just one shock angle α and associated Mach number M for which the Rankine-Hugoniot equations will be compatible with parallel flow upstream of the shock; we can write $\alpha = \alpha(M, \beta, \gamma)$. For given β, γ such a theoretical solution with "attached shock" (cf. §11) exists only for $M > M(\beta, \gamma)$. If $M < M(\beta, \gamma)$, no conical flow is possible, and a flow with "detached shock" is thus predicted theoretically.

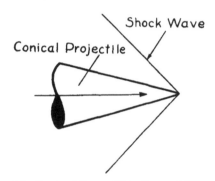

Fig. 24. Conical flow of Taylor and Maccoll

Related flows will be discussed in §88. Suffice it to say here that the predicted limits of the conical regime, the pressure on the cone head, and the angle of the attached shock wave, as functions of the Mach number and cone head angle, have been closely confirmed experimentally.

86. Expanding pressure waves

There also exist important families of non-stationary flows having the internal symmetry (18). Among such flow families the cases of expanding plane, cylindrical and spherical waves are especially noteworthy. Expanding plane waves arise, for example, if a diaphragm is ruptured in a shock tube, in the region behind a slab of explosive detonated from one face, or behind a piston given a constant impulsive velocity in an infinitely long cylinder.[15] Expanding spherical waves arise from a uniformly expanding sphere.

Expanding pressure waves are noteworthy as constituting one of the first conscious applications of the method of search for symmetric

[15] For expanding plane waves ("centered rarefaction waves"), see [2], Sec. 46. For the pressure wave due to an expanding sphere, see G. I. Taylor, *Proc. Roy. Soc.*, *A186* (1946), 273–92.

solutions.[16] We shall discuss them here from a mathematical standpoint only.

It is convenient to use Lagrangian coordinates, letting a measure the integrated mass from a fixed material point (e.g. at a stationary center of symmetry). For *plane* waves, if we denote position by $x = f(a, t)$ and density by $\rho = \rho(a, t)$, then the Equation of Continuity is equivalent to the relation $\sigma = \partial f / \partial a$ between the specific volume $\sigma = 1/\rho$, x, and a. Therefore, possible flows for a given Equation of State $p = p_0 - F(\sigma) = p_0 + k\rho^\gamma$ correspond to solutions of the Equations of Motion. These reduce to

$$(26) \qquad \frac{\partial^2 f}{\partial t^2} = F'(\sigma) \frac{\partial^2 f}{\partial a^2} = k\gamma \left(\frac{\partial f}{\partial a} \right)^{-\gamma-1} \frac{\partial^2 f}{\partial a^2},$$

as in [2], Sec. 18. (It is easily verified that $u = \partial f / \partial t$ is the velocity, $\partial^2 f / \partial t^2$ the material acceleration, and that the right side represents $-\partial p / \rho \partial x$.)

Since velocity $u = \partial f / \partial t$, self-similarity under (18) is equivalent to $f(\alpha a, \alpha t) = \alpha f(a, t)$ for all $\alpha > 0$, and hence to $f(\alpha, \alpha t) = \alpha f(1, t) = \alpha g(t)$. Setting $\tau = t/a$, this gives

$$(26') \qquad x = f(a, t) = af\left(1, \frac{t}{a}\right) = ag(\tau), \qquad \tau = \frac{t}{a}.$$

That is, (26') expresses invariance under (18).

Substituting (26') into (26), we get

$$(27) \qquad 0 = a^{-1}g''(\tau)\{1 - \gamma k\rho^{\gamma+1}\tau^2\},$$

since $\partial^2 f / \partial a^2 = t^2 g''(t/a)/a^3$ by direct calculation. Hence "centered" plane waves having the expansive symmetry (18) are the solutions of the ordinary differential equation (27). (Earnshaw's Paradox states that there are none having translation symmetry.) Equation (27) has two families of solutions. If $g'' = 0$, then $f = a[C_1 + C_2(t/a)] = C_1 a + C_2 t$. This implies uniform flow with constant u, σ and is trivial.

Otherwise, $1 = \gamma k\rho^{\gamma+1}\tau^2$, which implies

$$(28) \qquad \left(\frac{\partial f}{\partial a} \right)^{\gamma+1} = \sigma^{\gamma+1} = \gamma k\tau^2.$$

But by (26'), $\partial f / \partial a = g(t/a) - (t/a)g'(t/a)$; hence the condition is

$$(29) \qquad g - \tau g' = [\gamma k\tau^2]^{1/(\gamma+1)}, \quad \text{unless} \quad \gamma = -1.$$

This first-order linear non-homogeneous ordinary differential equation can easily be integrated formally. The general solution is

$$(29') \qquad g(\tau) = C\tau + A\tau^{2/(\gamma+1)},$$

[16] See [58], [69], and [71]. They are the theme of [56], Ch. IV. In [56], Ch. II, Sec. 13, reference is made to an earlier discussion of self-similar gravity waves by N. E. Kochin, *Trudy Steklov Inst.*, *9* (1935). See also [7], p. 385.

where $A = [(\gamma+1)/(\gamma-1)][\gamma k]^{1/(\gamma+1)}$, and C is arbitrary—provided $|\gamma| \neq 1$. There is no solution if $\gamma = -1$, since then $\tau = $ const. by (28). If $\gamma = 1$, the general solution is $g(\tau) = C\tau - \sqrt{\gamma k}\tau \ln \tau$.

The cases of centered cylindrical and spherical waves can be treated in the same way. In $m+1$ dimensions the Equations of Motion are

$$(30) \qquad \frac{\partial^2 r}{\partial t^2} = r^m F'(\sigma) \frac{\partial}{\partial a}\left[r^m \frac{\partial r}{\partial a}\right], \quad \sigma = r^m \frac{\partial r}{\partial a}.$$

The condition of self-similarity under (18) is equivalent to the following analog of (26'):

$$(30') \qquad r = bg(\tau), \quad \tau = t/b, \quad b = a^{1/(m+1)}.$$

Substituting (30') into (30) we again get a second-order ordinary differential equation whose solutions represent centered cylindrical and spherical waves.

As in §§84–85, analogs of the preceding waves can also be obtained for general Equations of State without the requirement that the gas be "polytropic."

87. Polytropic symmetry

In the polytropic case $p - p_0 = k\rho^\gamma$ (cf. Ch. IV, Theorem 9) the equations of compressible, nonviscous, homentropic flow admit a two-parameter group of symmetry. This is a subgroup of the three-parameter group of transformations

$$(31) \qquad \begin{aligned} x &\to \alpha x, \quad t \to \beta t, \quad u \to (\alpha/\beta)u, \\ \rho &\to \delta\rho, \quad (p-p_0) \to \delta^\gamma(p-p_0). \end{aligned}$$

Every transformation (31) leaves invariant the polytropic equation of state, and the equation of continuity $\partial\rho/\partial t + \mathrm{div}\,(\rho u) = 0$. It leaves invariant the equations of (nonviscous) motion if and only if $\delta^{\gamma-1} = \alpha^2/\beta^2$. Hence, the two-parameter subgroup of the group (31) which respects Euler's Equations of Motion is defined by $\delta = (\alpha/\beta)^{2/(\gamma-1)}$.

In any one-parameter subgroup of (31), except in the trivial case $\beta \equiv 1$, we have $\alpha = \beta^\tau$ for some constant exponent τ. Hence $\delta = \beta^{2(\tau-1)/(\gamma-1)}$ if the subgroup respects Euler's Equations of Motion, giving

$$(32) \qquad \begin{aligned} x &\to \beta^\tau x, \quad t \to \beta t, \quad u \to \beta^{\tau-1}u \\ \rho &\to \beta^{2(\tau-1)/(\gamma-1)}\rho, \quad (p-p_0) \to \beta^{2\gamma(\tau-1)/(\gamma-1)}(p-p_0). \end{aligned}$$

The self-similar flows of §§84–86 correspond to the choice $\tau = 1$, a case in which the second line of (32) simplifies to $\rho \to \rho$, $p \to p$, making (32) degenerate to (18).

The orbits of the group (32) (its "sets of transitivity") are, in space-time, the curves of constant $\chi = x/t^\tau$. Hence, the nonviscous compressible flows carried into themselves by the special group (32) are defined by the relations

$$(33) \qquad u_i(x; t) = t^{\tau-1} f_i(\chi), \qquad \rho = t^{2(\tau-1)/(\gamma-1)} /e(x)$$

as well as $p - p_0 = k\rho^\gamma$. Substituting into the Equations of Motion, we get necessary and sufficient conditions for a flow to be self-similar under the special group of Mach modeling.

An important example of such a flow describes the asymptotic expansion of explosives gases in the bore of a cannon while accelerating a projectile of constant inertia.[17] The Equation of Continuity can be eliminated by using Lagrangian coordinates. Moreover, there is a singular "point concentration" of initial energy at $t=0$, much as in the first example of §80. This corresponds to the case of a highly concentrated explosive in a "long-bore" cannon; *adiabatic* flow may be assumed.

A related example concerns extremely intense spherical and cylindrical blast waves so intense that pressure outside the blast can be neglected.[18] In this case the entropy depends on the strength of the shock wave and decreases with time; $\tau = 2/5$ to conserve total energy.

For the final formulas in these cases, the reader may consult the references given above.

88. Conical flows

The flows described so far have sufficient physical symmetry in space-time, so that all quantities involved could be expressed as functions of *one* independent variable in each case. In such cases, the partial differential equations of fluid mechanics reduce to ordinary differential equations. However, there are other important applications of the method of search for symmetric solutions leading to *partial* differential equations. We shall now consider some such applications.

The most obvious case concerns so-called "conical flows" without axial symmetry. As first defined and considered by A. Busemann,[19] these are steady flows whose velocity fields satisfy

$$(34) \qquad u = u(\phi, \theta)$$

---

[17] A. E. H. Love and F. B. Pidduck, *Phil. Trans.*, *A222* (1922), 167–226; R. H. Kent, *Physics*, 7 (1926), 319–24. The acceleration $a \propto t^{\tau-2}$. See also [2], Sec. 160.

[18] See [2], Sec. 161; the model is due to G. I. Taylor, *Proc. Roy. Soc.*, *A201* (1950), 159–86. For further results see [56], Ch. IV.

[19] See *NACA Tech. Memo. 1100* (1947), and refs. given there. Also [10], Sec. 10.5; A. Ferri in Ch. H3 of *General theory of High Speed Aerodynamics*, Princeton Univ. Press, 1954.

in spherical coordinates. Such flows occur with delta wings, for example, because delta wings have conical symmetry.

A more subtle application, to expanding self-similar flow, concerns the entry at constant speed of a wedge or cone into water (see Fig. 25), at sufficiently high speeds so that inertial forces predominate during entry. We consider first the case of a wedge.

As before, the transformation

$$(35) \qquad x_i \rightarrow \alpha x_i, \qquad t \rightarrow \alpha t, \qquad p, \rho, u_i \text{ unchanged},$$

leaves inertial fluid mechanics unchanged—indeed, we could suppose the

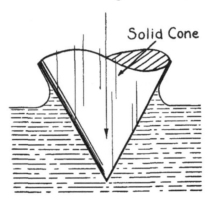

Fig. 25. Vertical impact of cone on water

fluid compressible! Hence the method of "search for symmetric solutions" leads one to look for solutions of the form

$$(36) \qquad U(x, y; t) = t\phi\left(\frac{x}{t}, \frac{y}{t}\right),$$

in the case of wedges. This method of reduction from three to two independent variables is the basis of a well-known paper of H. Wagner on the impact of seaplane floats.[20] We can reduce to functions of a *single* complex variable by the considerations of Chapter III, §2, but the boundary conditions are complicated.

The same method is evidently applicable to a cone entering the water at a constant velocity, suggesting

$$(37) \qquad U(x, y, z; t) = t\phi\left(\frac{x}{t}, \frac{y}{t}, \frac{z}{t}\right),$$

i.e. a reduction from *four* to *three* independent variables. In the case of a

[20] H. Wagner, *Zeits. ang. Math. Mech.*, *12* (1932), 193–215.

right *circular* cone entering the water vertically, solutions having the circular symmetry of the hypotheses can be expressed through

$$U(x, y, \theta; t) = t\phi\left(\frac{x}{t}, \frac{y}{t}\right),$$

in terms of a single function of *two* independent variables.

In the incompressible case, one can use potential theory to generate the flow-field by a *source distribution* on the free surface, whose location and intensity are unknown functions of a single variable (arc-length). This idea was applied by Shiffman and Spencer,[21] who showed that the "free-surface" condition of constant pressure led to a system of integral equations in functions of *one* variable. The approximate numerical integration of these, for a 60° cone, was an important achievement of Hillman.

89. Local and global solutions

The above examples show that self-similar mathematical solutions exist, in many cases, to problems having a given symmetry in space-time. The formulation and proof of general existence theorems is, however, much more difficult.

In cases involving sufficient symmetry, so that the usual differential equations of fluid flow reduce to *ordinary* differential equations, standard *local* existence theorems are available. However, the existence of *global* solutions, satisfying plausible boundary conditions as well, is more unpredictable. A clear instance of this unpredictability is furnished by compressible nonviscous plane flow having rotational symmetry (spiral streamlines). As first shown by Ringleb,[21a] this flow is impossible in the large, since the radial flow component reverses direction along a "limit circle".

This unpredictability is also graphically illustrated by the case of Taylor-Maccoll flows (§85), for which the "attached shock" regime of conical symmetry is limited to sufficiently small cone head angles for each Mach number. The general class of steady axially symmetric conical flows satisfying (25) is very hard to determine rigorously in the large, and published statements are not always reliable.[22]

Similarly, although shock-free centered rarefaction waves are possible, compression waves involve shocks, so that discussions in the large of self-similar blast waves are quite complicated.

---

[21] *Comm. Pure Appl. Math.*, *4* (1951), 379–417, where Hillman's results are also reported. See also [15], Ch. XI, Sec. 9.

[21a] *ZaMM*, *20* (1940), 185–98. See also [5], Chs. V. 4 and VII. 8.

[22] The most thorough study is that of the author and J. M. Walsh, *Riabouchinsky Jubilee Volume*, Paris, 1954, 1–12.

Another interesting illustration of global difficulties is furnished by (laminar, viscous) axially symmetric *jets*. As shown in §83, one can reduce the Navier-Stokes equations to ordinary differential equations by postulating "self-symmetric" velocity-fields of the form

$$(38) \qquad \qquad u = r^{-1}f(\theta),$$

in spherical coordinates. Unfortunately, as shown by Beran,[22a] the resulting ordinary differential equation (17) has no global solutions which satisfy the natural boundary condition for the jet from a circular hole in a plane wall, or any other conical orifice. Contrary to some published statements, only the jet from a tube with parallel walls seems to be mathematically compatible, in the large, with the symmetry (38) postulated and natural boundary conditions.

Local existence theorem. Even general *local* existence theorems are not easy to prove. One positive result is the following:[23] (The reader's indulgence is asked for the abstract mathematical terminology which is used in the interest of brevity.)

Theorem 1. Let $X = \Gamma \times E$ be the direct product of its subspaces Γ and E, and for each fixed $\alpha \in E$ let the group G of transformations of X be transitive [23a] on the set of (γ, α), for variable $\gamma \in \Gamma$. If G leaves invariant a differential equation $D[u] = 0$ defined in X, then there exists a differential equation $\Delta[U] = 0$ on E of at most the same order, such that $u(x) = u(\gamma, \xi) = U(\xi)$ satisfies $D[u] = 0$, if and only if $U(\xi)$ satisfies $\Delta[U] = 0$ for $\xi \in E$.

Proof. Near each point $x = (\gamma, \xi)$ of X, we can set up local coordinates $\gamma_1, \cdots, \gamma_r$ and $\xi_1, \cdots, \xi_{n-r}$ in X. Any p-th partial derivative $X^{(p)}[u]$ with respect to these coordinates will have the simple form $\Gamma^{(m)}[u]E^{(p-m)}[u]$, where $\Gamma^{(m)}$ and $E^{(p-m)}$ are partial derivatives with respect to the coordinates $\gamma_1, \cdots, \gamma_r$ and $\xi_1, \cdots, \xi_{n-r}$ for Γ and E, respectively. Hence any partial differential operator $D = \Phi\{X_1^{(p_1)}, \cdots, X_s^{(p_s)}\}$ of order q on functions $u(x)$ defined on X can be written in a form

$$(39) \qquad D = \Psi\{\Gamma_1^{(m_1)}, \cdots, \Gamma_s^{(m_s)}; \quad E_1^{(p_1-m_1)}, \cdots, E_s^{(p_s-m_s)}\},$$

which involves a function of partial derivatives of order at most q on E.

But for functions $U(\xi) = u(\gamma, \xi)$ whose value at any point $x = (\gamma, \xi)$ depends only on γ (i.e., for functions invariant under G), the effect of any operator E_j is to transform them into another function of the same class, while (G being transitive) the effect of any operator Γ_j is to give the

[22a] M. Beran, *Quar. Appl. Math.*, **14** (1956), 213–14.

[23] See also J. A. Morgan, *Quar. J. Math.*, **3** (1952), 250–9. With linear differential equations and compact G, group integration gives another approach.

[23a] This means that, given (γ, α) and (γ', α), g exists in G such that $g(\gamma, \alpha) = (\gamma', \alpha)$. We assume that Γ and E are differentiable manifolds.

function 0. Hence, as regards such functions, D is equivalent to the differential operator on E obtained by suppressing all terms for which $m_j > 0$. This proves the theorem.

Corollary. If $D[u]=0$ is well-set, for a class of boundary conditions invariant under G, then $\Delta[u]=0$ is well-set.

Whereas analyticity is not essential for proving local existence theorems on ordinary differential equations,[24] it is often an essential hypothesis for existence theorems on partial differential equations. In the case of *analytic* partial differential equations (and analytic groups of symmetry), the reduced equation (39) will also be analytic. In this case many *initial-value* problems will at least satisfy *local* existence theorems. For, suppose that all *time* derivatives of functions $\phi_i(x; t)$, $x=(x_1, \cdots, x_n)$ involved can be expressed in terms of the ϕ_i and their first *space* derivatives, so that, symbolically,

$$(40) \qquad \frac{\partial \phi_i}{\partial t} = F_i\left(\phi_j, \frac{\partial \phi_j}{\partial x_k}\right).$$

Then the Cauchy-Kowalewski existence theorem[25] asserts that (40) has one and only one local analytic solution for given analytic initial conditions $\phi_i(x; 0)$ at $t=0$.

But now suppose that (40) is invariant under a group G. Let $\phi_i(x; 0) = G_i(x)$ be a set of analytic initial conditions invariant under G. Then the *unique* solution defined locally by the preceding theorem will also be invariant under G. Hence we get a local *existence* (and uniqueness) theorem for the reduced differential equation, obtained by the method of search for symmetric solutions, whenever we have one for the original differential equations.

The seemingly innocuous restriction to *first* space derivatives in (40) is really quite strong. Thus, it implies that the system (40) must be of hyperbolic type. It is fulfilled in a compressible nonviscous fluid but not, for example, in an incompressible nonviscous fluid or in any viscous fluid. To establish rigorously even the local validity of the method of search for symmetric solutions will require much further research into partial differential equations.

90. Groups and separable variables

Solutions of physical problems possessing internal symmetry with respect to a group can usually be given a simplified mathematical form by the introduction of suitable variables associated with this group. I shall now show how such substitutions lead to "separations of variables" of great use in fluid mechanics.

---

[24] I have not studied conditions required to rule out difficulties which may arise for ordinary differential equations like $y'^2 + y^2 + 1 = 0$, of "degree" greater than one.

[25] J. Hadamard, *Le problème de Cauchy*, Paris, 1932, Ch. I.

For example, let us reconsider the invariance of the Euler-Lagrange equations for a nonviscous compressible fluid, under the group

(18) $$t \to \alpha t, \qquad x_i \to \alpha x_i, \qquad p, \rho, u \text{ fixed.}$$

By definition, individual "self-symmetric" flows, invariant under the group (18), can be expressed in the form

(41) $$u_i = f_i(\mathbf{X}), \qquad p = p(\mathbf{X}), \qquad \rho = \rho(p) = \rho(p(\mathbf{X})),$$

where

(42) $$\mathbf{X} = (\mathbf{X}_1, \mathbf{X}_2, \mathbf{X}_3) = \left(\frac{x_1}{t}, \frac{x_2}{t}, \frac{x_3}{t}\right) = \frac{\boldsymbol{x}}{t}.$$

Clearly, (42) is a special case of

(43) $$\mathbf{X} = h(t)\boldsymbol{x}.$$

We will now find all unsteady flows of a nonviscous fluid which admit the formal "separation of variables" (41) and (43).

Our first result will be negative. It will be that, in this case, every such flow is invariant under (18): nothing is gained by generalizing (42) to (43).

For any differentiable function $F(\mathbf{X})$, (43) clearly implies

(44) $$\frac{\partial F}{\partial t} = \frac{h'}{h} \sum \mathbf{X}_k \frac{\partial F}{\partial \mathbf{X}_k} \quad \text{and} \quad \frac{\partial F}{\partial x_i} = h \frac{\partial F}{\partial \mathbf{X}_i},$$

summation being over k. Hence the equation of continuity, $\partial\rho/\partial t + \text{div}(\rho u) = 0$, is equivalent to

(45) $$h\left[\left(\frac{h'}{h^2}\right) \sum \frac{\mathbf{X}_k \partial \rho}{\partial \mathbf{X}_k} + \sum \frac{\partial(\rho u_k)}{\partial \mathbf{X}_k}\right] = 0,$$

if (43) holds. Likewise, the equations of motion for a nonviscous fluid are equivalent, gravity being neglected, to

(46) $$h\left[\left(\frac{h'}{h^2}\right) \sum \frac{\mathbf{X}_k \partial u_i}{\partial \mathbf{X}_k} + \sum \frac{u_k \partial u_i}{\partial \mathbf{X}_k} + \frac{\partial p}{\rho \partial \mathbf{X}_i}\right] = 0.$$

If h'/h^2 is a non-zero constant in time, say $-C$, then $1/h = C(t-t_0)$. Hence, after an obvious change of time origin and unit, we can reduce to the case of flows satisfying (42), and hence possessing internal symmetry under the group (18).

Otherwise $\sum \mathbf{X}_k \partial\rho/\partial \mathbf{X}_k = \sum \mathbf{X}_k \partial u_i/\partial \mathbf{X}_k = 0$, as can be shown by differentiating (45)–(46) partially with respect to time for fixed $\mathbf{X}$. In this case, by (44), $\partial\rho/\partial t = \partial u/\partial t = 0$ and so $\rho = \rho(\boldsymbol{x})$, $u = u(\boldsymbol{x})$. Hence, the exceptional case of variable h'/h^2 in (45) gives just the steady *conical* flows of §88. Such flows satisfy (41) and (43) for *any* $h(t)$, in particular for $h(t) = 1/t$ as in (42); they are all self-similar under (18).

In summary, the method of "search for symmetric solutions" under the group (18) gives *all* the nonviscous flows admitting the (apparently more general) "separation of variables" (41) and (43).

In the *irrotational* case, (41) and (42) are equivalent to the assumption that the velocity potential $U(x; t)$ admits the separation of variables

$$(47) \qquad\qquad U = tF(X) = tF(x/t)$$

already made in (36) and (37). For irrotational *homentropic* flows one can then apply the generalized Bernoulli equation of §4, $\partial U/\partial t + \frac{1}{2}\nabla U \nabla U + \int dp/\rho = C(t)$. This reduces to

$$(47') \qquad\qquad F(X) - \sum X_k \frac{\partial F}{\partial X_k} + \frac{1}{2}\nabla F \nabla F + \int \frac{dp}{\rho} = C,$$

by (44).

In the *incompressible* case $\rho = \rho_0$, an extended class of similarity solutions can be obtained by letting $C = C(t)$ be variable.

Further generalizations. The separation of variables (47), though equivalent to (41), suggests formally that one should consider more generally all flows which are *self-similar* in time, in the sense that

$$(48) \qquad\qquad u = g(t)f(X), \quad \text{where} \quad X = h(t)x.$$

This class of flows includes the flows treated in §87, for which (as noted after (32)) invariance under (18) is equivalent to $\tau = 1$.

It also includes a novel class of *incompressible* irrotational flows with free boundaries, due to von Karman.[26] These are defined by the similarity condition

$$(49) \qquad\qquad U(x; t) = \frac{1}{t}\,\phi(x),$$

corresponding to a constant acceleration coefficient.

91. Viscous case

It would be very interesting to determine the most general nonviscous flow satisfying the similarity condition (48), and to test such flows for invariance under subgroups of the dimensional group. Instead, to maintain balance, we shall determine the *incompressible, viscous* flows which satisfy (48).

As in §3, the equations of state and continuity are together equivalent in incompressible flow to the single condition div $u = 0$. Since g and h are non-vanishing, this is equivalent to

$$(50) \qquad\qquad \sum \frac{\partial f_k}{\partial X_k} = 0,$$

from which t has been eliminated.

[26] *Annali di Mat.*, 29 (1949), 247–49. See also [15], p. 248.

It remains to consider the Navier-Stokes equations of motion. By Theorem 1 of §21 we can neglect gravity. The other terms are, by direct substitution from (48) in the Navier-Stokes equations of Chapter II, (3),

$$\frac{\partial u_i}{\partial t} = g'f_i + \sum gh'x_k \left(\frac{\partial f_i}{\partial X_k}\right) = g'f_i + \sum \left(\frac{gh'}{h}\right) X_k \frac{\partial f_i}{\partial X_k},$$

$$\sum u_k \frac{\partial u_i}{\partial x_k} = \sum gf_k gh \left(\frac{\partial f_i}{\partial X_k}\right) = \sum (g^2 h) f_k \frac{\partial f_i}{\partial X_k},$$

$$\frac{1}{\rho} \frac{\partial p}{\partial x_i} = \left(\frac{a(t)h(t)}{\rho_0}\right) \frac{\partial p}{\partial X_i} \quad \text{if} \quad p = a(t)p(X),$$

$$\nu \nabla^2 u_i = \nu(gh^2) \sum_k \frac{\partial^2 f_i}{\partial X_k^2}.$$

Hence the Navier-Stokes differential equations can be expressed in the "separated form"

$$(51) \qquad \sum_{j=1}^{5} F_j(t) G_j(X) = 0,$$

where $F_1 = g'$, $F_2 = gh'/h$, $F_3 = g^2h$, $F_4 = ah$, $F_5 = gh^2$, and

$$G_1 = f_i, \qquad G_2 = X_k \frac{\partial f_i}{\partial X_k}, \qquad G_3 = f_k \frac{\partial f_i}{\partial X_k},$$

$$G_4 = \frac{1}{\rho_0} \frac{\partial p}{\partial X_i}, \quad \text{and} \quad G_5 = -\nu \sum \frac{\partial^2 f_i}{\partial X_k^2}.$$

Clearly condition (51) is equivalent to the requirement that the vectors $F = (F_1, F_2, F_3, F_4, F_5)$ and $G = (G_1, G_2, G_3, G_4, G_5)$ be confined to *orthogonal subspaces*. Depending on the number of linearly independent relations satisfied by the G_j, the "F-subspace" spanned by the F_i may conceivably have 1, 2, 3, or 4 dimensions. I shall first investigate the non-degenerate case, when all the F_j are proportional, so that the F-subspace has one dimension.

We can make F_4 proportional to F_3, by making $a = \alpha g^2$. Again, $g = h$ is (we can ignore constant factors) equivalent to the proportionality of F_1 and F_2. The remaining condition is that F_1 and F_3 be proportional, or that $g' = (-\beta/2)g^2h$ for some constant β. Since $g = h$, this amounts to $-2g'/g^3 = \beta$, or $1/g^2 = \beta(t - t_0)$. By properly choosing the origin and scale of time, this can be reduced to $g = 1/\sqrt{t}$, and hence to

$$(52) \qquad u_i(x; t) = \frac{1}{\sqrt{t}} f_i(X) \quad \text{and} \quad p = \frac{p(X)}{t} \left[X = \frac{x}{\sqrt{t}}\right],$$

$$(52') \qquad -\frac{1}{2}\left(f_i + X_k \frac{\partial f_i}{\partial X_k}\right) + f_k \frac{\partial f_i}{\partial X_k} + \frac{1}{\rho_0} \frac{\partial p}{\partial X_i} = \nu \sum_k \frac{\partial^2 f_i}{\partial X_k^2}.$$

Hence we have again reduced the number of independent variables by one—by a search for a more general type of symmetric solutions! I omit the discussion of (52′) itself.[27]

This corresponds to the invariance of the Navier-Stokes equations under $t \to \alpha t$ and $\boldsymbol{x} \to \beta \boldsymbol{x}$, provided the Reynolds number $R = vd/\nu \propto (\beta/\alpha)\beta$ [$\nu = $ const.] is unchanged—so that $\beta \propto \sqrt{\alpha}$. The solutions (52) are precisely the flows invariant under this group.

There are also "degenerate" solutions of (51). For example, consider the "parallel flows" parallel to the x-axis, with

$$(53) \qquad u_1 = g(t)f_1(y, z), \qquad u_2 = u_3 = p = 0.$$

In this case (50) is always satisfied. Since $h = 1$, $F_2 = 0$; since $p = 0$, $F_4 = 0$. What matters more, $G_4 = f_1 \partial u_i/dx = 0$ for all i; hence F_3 is unrestricted. It remains to satisfy $g'f_1 = \nu g(\partial^2 f_1/\partial y^2 + \partial^2 f_1/\partial z^2)$, which, g being a function of t and f_1 of $\boldsymbol{x} = (y, z)$, reduces to $g'/g = -k$ and

$$(53') \qquad u_1 = e^{-kt}f_1(y, z), \quad \text{where} \quad \frac{\partial^2 f_1}{\partial y^2} + \frac{\partial^2 f_1}{\partial z^2} = -\nu k f_1.$$

This defines the well-known[28] exponential decay of parallel viscous flows, for example of flow in the two-dimensional channel $-a \leqq y \leqq a$, with $u_1 = e^{-kt} \cos \pi y/2a$, and $k = \pi^2/4a^2\nu$.

92. Inverse methods

The preceding examples reveal the method of "separation of variables" as a generalization of the method of "search for symmetric solutions". In turn, the method of separation of variables is a special case of a wide class of "inverse methods" discussed systematically by the late P. Nemenyi.[29] The relation may be described loosely as follows. Whenever group theory indicates the existence of flows with separated variables, or any other property P, an *a priori* postulation of property P will yield these solutions at least, and possibly others.

For example, group theory assures use of the existence (locally) of Prandtl-Meyer expansion waves, in which

(P1) the vector velocity is constant along each straight line of a one-parameter family.

[27] In the plane, (50) is equivalent to the existence of a stream function, and p can be eliminated from (52′) by using curl (grad p) = 0. This leaves the fourth-order partial differential equation (6) invariant.

[28] See G. I. Taylor, *Phil. Mag.*, *46* (1923), 671–674. A similar exponential decay of circular flows is possible, with $p'(r)$ sufficient for centripetal acceleration. Cf. [68], and Ratip Berker, *Sur quelques cas d'intégration des équations du mouvement d'un fluide visqueux incompressible*, Lille, 1936.

[29] P. Nemenyi, *Advances in Applied Mechanics*, 2 (1951), 123–53.

The "inverse method" consists in finding *all* stationary irrotational flows of a compressible nonviscous fluid which have property P1. This can be done as follows.

We know (§5) that with stationary irrotational flow, the equations of motion are equivalent to the Bernoulli equation $u^2/2 + \int dp/\rho = C$. Hence, by numerical integration, there is for each "stagnation pressure" (i.e. constant of integration) one and only one pair of functions $\rho(u)$ and $p(\rho(u))$ such that the equations of state and motion are both satisfied. Moreover the flows with property P1 are the flows with such $\rho(u)$ and $p(\rho(u))$, which are *irrotational* and satisfy the *equation of continuity* (mass-conservation).

In the present problem, considerable geometrical insight can be obtained by using special coordinates appropriate to a one-parameter family of straight lines. These special coordinates are: the angle θ made

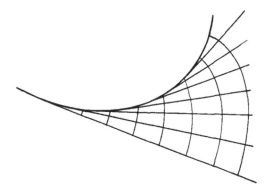

FIG. 26. Coordinates for Prandtl-Meyer expansion fan

with the x-axis by the lines, and the directed distance h from a fixed curve cutting the lines orthogonally, as in Fig. 26. If we recall that the lines represent, "in general," the tangents of a suitable plane curve Γ, we will see immediately: (i) the loci $\theta = $ const. are the given straight lines, (ii) the lines $h = $ const. are the orthogonal family of "evolutes" of Γ, (iii) $ds^2 = dh^2 + r^2 d\theta^2$, where $r = h + s(\theta)$ is the radius of curvature of the evolute, and s denotes arc-length along Γ.

In terms of these intrinsic geometrical coordinates, we can easily express the two conditions of irrotationality and mass conservation.

By definition, irrotationality is the condition that the circulation around any closed curve γ vanishes. If α denotes the angle from the straight line $\theta = \theta_0$ to the velocity vector $u(\theta_0)$, of magnitude q, the circulation around γ is

$$\oint \sum u_k dx_k = \oint q(\cos \alpha \, dh + \sin \alpha \, r d\theta).$$

By Green's Theorem, this vanishes identically if and only if

$$\frac{\partial}{\partial \theta} (q \cos \alpha) = \frac{\partial}{\partial h} (rq \sin \alpha).$$

Since $r = h + s(\theta)$, $\partial r / \partial h = 1$; while by P1, q and α are functions of θ alone. Hence irrotationality is equivalent to

(54) $$\frac{d}{d\theta} (q \cos \alpha) = q \sin \alpha.$$

Letting $g(\theta) = q \cos \alpha$, and $h(\theta) = q \sin \alpha$, this is $g' = h$, generalizing (20) of §84.

To express the condition of mass conservation, observe that the rate of mass flux *outward* through a curve γ is

$$\oint_\gamma \rho q(-\sin \alpha dh + \cos \alpha \, rd\theta).$$

By Green's Theorem, this vanishes identically if and only if

(55) $$\frac{\partial}{\partial \theta} (-\rho q \sin \alpha) = \frac{\partial}{\partial h} (r\rho q \cos \alpha) = \rho q \cos \alpha.$$

In the notation just introduced, this reduces to the equation $(-\rho h)' = \rho g$, where the accent denotes differentiation with respect to θ. Substituting $h = g'$ and simplifying, this becomes $\rho' g' + \rho g'' + \rho g = 0$, or $(g'' + g) + (\rho'/\rho)g' = 0$, which is (22) of §84.

Hence *all flows satisfying condition P1 can be obtained from Prandtl-Meyer flows*, by replacing the radii from a fixed corner by the tangents to a fixed curve Γ, and using the same vector velocity at corresponding points.[30]

The preceding illustration is a special case of the more general "inverse problem" of finding all flows with one-dimensional hodographs—that is, all flows whose velocity vectors are confined to lie on a single *curve*.[31] (In general, the locus of all velocity vectors of a flow is called its "hodograph".)

93. Discussion

Obviously, the method of search for symmetric solutions is just one method for postulating arbitrary functional relationships and searching

---

[30] This result is due to F. H. Lees, *Proc. Camb. Phil. Soc.*, 22 (1924), 350–62; see also [2], pp. 273–278, and refs. given there.

[31] For the incompressible, viscous case, see W. Müller, *Einführung in die Theorie der zähen Flüssigheiten*, Leipzig, 1932; also *Zeits. ang. Math. Mech.*, 13 (1938), 395–408. For the compressible, nonviscous case, see [69]; also J. H. Giese, *Quar. Appl. Math.*, 9 (1951), 237–46.

for flows satisfying these relations. The method of separation of variables is another such method. Thus, the class of "inverse methods" includes the method of search for symmetric solutions *and* the method of separation of variables as special cases.

The great advantage of the search for symmetric solutions over the other two' methods is the availability of theorems guaranteeing the existence of symmetric solutions, at least locally (cf. §89). In fact, when separation of variables is compatible with nontrivial solutions, they are usually associated with groups.

This principle is illustrated by the Laplace equation $\nabla^2 U = 0$ for steady Euler flows in space and the Helmholtz equation $\nabla^2 U + k^2 U = 0$. In these cases, it has been shown[31a] that the coordinate systems compatible with separations of variables belong to a few familiar classes, of which most are equivalent under the group generated by inversions in spheres to parallel planes, planes through a line, and concentric spheres—i.e. to one set of coordinate surfaces for Cartesian, cylindrical, or spherical coordinates. This suggests that the problem could be attacked directly by conformal methods invariant under the conformal group.

However, it would be an overstatement to claim that all of the separations of variables of fluid mechanics are correlated with groups (internal symmetries). Although von Karman's solution[32] of the Navier-Stokes equations for the flow near a rotating disc admits the affine internal symmetries

$$r \rightarrow \alpha r, \quad z \rightarrow z, \quad u_r \rightarrow \alpha u_r, \quad u_\theta \rightarrow \alpha u_\theta, \quad u_z \rightarrow u_z,$$

these do not leave the Navier-Stokes equations invariant.

Similarly, the "inverse" assumptions of constant velocity magnitude or vorticity on streamlines, etc., do not have any obvious relation to groups.[33] It would be desirable to know all the coordinate systems in which the solutions of the equations of time-dependent fluid motion can be found by separating variables, just as these are known for the Laplace and Helmholtz equations.[31a]

94. Hodograph method

A notable simplification of the equations of compressible nonviscous flow is provided by the hodograph transformation. We have already seen (Chapter I, Eq. (10)) that the steady irrotational plane flows of a

---

[31a] See [67]; also P. Moon and D. E. Spencer, *Proc. Am. Math. Soc.*, *3* (1952), 635–42 and *4* (1953), 302–7, and the refs. given there.

[32] Th. von Karman, *ZaMM*, *1* (1921), 233–52; G. Batchelor, *Quar. J. Math. Appl. Mech.*, *4* (1951), 29–41. For a further generalization, see R. Berker, *op. cit.* in ftnt. 28.

[33] J. Kampé de Fériet, *Proc. Int. Math. Congress Zürich* (1932), Vol. 2, 298–99. P. Nemenyi and R. Prim, *J. Math. Phys. MIT*, *27* (1948), 130–35.

compressible nonviscous fluid correspond one to one to the velocity potentials U which satisfy the *nonlinear* partial differential equation

$$(56) \qquad \nabla^2 U = \frac{1}{c^2} \left\{ U_x U_x U_{xx} + 2 U_x U_y U_{xy} + U_y U_y U_{yy} \right\}.$$

Here subscripts connote differentiation with respect to the variable involved, and c^2 is the local velocity of sound.

We recall [34] that (56) is equivalent to either of the *linear* partial differential equations

$$(57a) \qquad q^2 V_{qq} + q \left(1 + \frac{q^2}{c^2}\right) V_q + \left(1 - \frac{q^2}{c^2}\right) V_{\theta\theta} = 0,$$

or

$$(57b) \qquad q^2 \phi_{qq} + q \left(1 - \frac{q^2}{c^2}\right) \phi_q + \left(1 - \frac{q^2}{c^2}\right) \phi_{\theta\theta} = 0.$$

Here V is the stream function; $q e^{i\theta}$ is the complex velocity vector, so that $U_x = q \cos\theta$ and $U_y = q \sin\theta$; c^2 is a single-valued function of q, by the Bernoulli equation; and $\phi = U - x U_x - y U_y$ is the dependent variable of the Legendre contact transformation giving (57b).

Now it is not difficult to calculate (57a)–(57b) from (56) and the preceding definitions, but it is quite obscure why the hodograph variables leading to these linear equations should have been tried. One reason may have been the success of the hodograph method in free streamline problems (as in §38). I shall now give an alternative motivation in terms of three group-theoretic considerations.

The first group-theoretic consideration is the invariance of the laws of nonviscous fluid dynamics under the group (18) of Langevin transformations

$$x \to \lambda x, \qquad u \to u, \qquad U \to \lambda U, \qquad V \to \lambda V, \qquad \phi \to \lambda \phi.$$

It follows that a partial differential equation expressing these laws must ordinarily be *non-homogeneous* (and hence nonlinear) if the x_i are taken as the independent variables, and U, V, or ϕ as dependent variables— while it will be *homogeneous* if the u_i are taken as independent variables [35] and U, V, or ϕ as the dependent variables.

It is not clear to me *a priori* why this homogeneous equation should be linear when V and ϕ are taken as the dependent variables.

A second group-theoretic consideration is the obvious invariance of the laws of fluid motion under the group of *rotations* $\theta \to \theta + \alpha$, when q, U, V, ϕ

[34] See [2], Ch. IVA, or H. Seifert, *Math. Annalen*, 120 (1947), 75–126.

[35] This is possible locally except in the case, mentioned in §92, of a one-dimensional hodograph. It is not generally possible in the large, for reasons similar to those described in §89.

are held fixed. This implies that, in (57a)–(57b), θ will appear only in differential operators and not in coefficients. Hence we have a group-theoretic motivation for the use of q, θ as independent variables[36] in place of $u = U_x$ and $v = U_y$. It makes the coefficients of our differential equations functions of *one* of the two independent variables only.

A third group-theoretic observation is the obvious invariance of the laws of fluid motion under the group of *translations* $x \to x + a$, when u, U, V, ϕ are held fixed (§67). This is equivalent to the occurrence of x and y in (56) as differential operators alone and not in coefficients.

95. Inertial plane motion

Group theory can not only be used to simplify the equations of fluid motion; it can also be used to reduce their integration to quadratures.[36a] An important illustration of this principle is provided by the motion of a missile in a plane, subject to *inertial* forces alone. (This is approximately the case in many ballistic problems, and also in submarine motion with a fixed rudder setting, when hydrostatic buoyancy compensates for gravity.) That is, we assume the group of §70.

Let $x = q_1$ and $y = q_2$ denote the coordinates of the missile, and $\psi = q_3$ the angle between the x-axis and an axis fixed in the missile. We *assume* that the instantaneous coordinates of position of the missile determine its future trajectory through force-reactions and Newton's laws, so that (subject to the usual limitations on differentiability, etc.)

$$(58) \qquad \ddot{q}_i = F_i(q_1, q_2, q_3; \dot{q}_1, \dot{q}_2, \dot{q}_3) = F_i(q; \dot{q}).$$

This is evidently a *sixth* order system of ordinary differential equations, which are in general nonlinear.

I shall now show how it can be reduced to a *second* order system and four quadratures by use of group theory. The method indicated generalizes in principle the usual method of "ignorable coordinates",[37] from Lagrangian to non-Lagrangian dynamical systems. After describing the reductions I shall sketch the further extension to general ordinary differential equations invariant under a group.

First, the system (58) may be expected to be invariant under *translations* $x \to x + x_0$, $y \to y + y_0$ of space-coordinates.[38] This has a very simple

---

[36] The use of $w = \ln q$ instead of q is suggested by complex variable theory: $w + i\theta = \ln (u + iv)$.

[36a] For a rigorous modern mathematical discussion of this classic relation, in the case of (ordinary) homogeneous *linear* differential equations, see E. R. Kolchin, *Annals of Math., 49* (1948), 1–42. For the nonlinear case, see Dickson [62].

[37] As described for example in [73], p. 54. With non-zero C_D, the system discussed here is non-Lagrangian.

[38] In practice this means that such factors as the variation of density with altitude can be neglected.

mathematical interpretation; it permits us to replace (58) by the *fourth order* system

(59)

$$\frac{d\dot{q}_1}{dt} = F_1(q_3; \dot{q}), \qquad \frac{d\dot{q}_2}{dt} = F_2(q_3; \dot{q}),$$

$$\frac{dq_3}{dt} = \dot{q}_3, \qquad \frac{d\dot{q}_3}{dt} = F_3(q_3; \dot{q}),$$

and the two quadratures

(59′) $$q_1 = \int \dot{q}_1 \, dt, \qquad q_2 = \int \dot{q}_2 \, dt.$$

Thus a two-parameter group permits us to reduce the order of our system by two, replacing integrations by quadratures. Second, the system (58) is "isotropic," that is, invariant under *rotations* of coordinates. To express this fact analytically it is convenient to use as new variables the scalar velocity $v = (\dot{x}^2 + \dot{y}^2)^{1/2}$, and the inclination θ of the trajectory to the x-axis. This makes $\dot{x} = v \cos \theta$ and $\dot{y} = v \sin \theta$. Clearly v and the "pitch angle" $\phi = \psi - \theta$ are invariant under rotations; hence (59) is equivalent [39] (if isotropic) to a system

(60)

$$\frac{dv}{dt} = G_1(v, \phi, \dot{\phi}), \qquad \frac{d\theta}{dt} = G_2(v, \phi, \dot{\phi}),$$

$$\frac{d\phi}{dt} = \dot{\phi}, \qquad \frac{d\dot{\phi}}{dt} = G_3(v, \phi, \dot{\phi}),$$

the second of which reduces to the quadrature $\theta = \int \dot{\theta} dt$.

Finally, the hypothesis that all forces are "inertial" means that they are proportional to velocity squared, i.e. that space trajectories are invariant under the group of changes in the time scale. But clearly ϕ and *distance* $s = \int v \, dt$ are invariant under this group. Hence, if we replace t by the independent variable s, we will have

(61) $$\frac{d\phi}{ds} = \phi', \qquad \frac{d\phi'}{ds} = H(\phi, \phi'), \qquad \frac{dv}{v \, ds} = H^*(\phi, \phi').$$

(For example, $m v \, dv/ds$ is the tangent force-component; this is v^2 times the force $m G_1(v/v, \phi, \dot{\phi}/v) = m G_1(1, \phi, \phi') = H^*(\phi, \phi')$ which would act if all velocities were reduced in the ratio $v:1$.) In summary, (58) *is equivalent to the second order system* (61), *plus the five quadratures* (59′) *and*

(62) $$v = v_0 e^{\int H^* \, ds}, \qquad t = \int \frac{ds}{v}, \qquad \theta = \int \dot{\theta} \, dt.$$

---

[39] Strictly speaking, as long as v does not vanish.

96. Theorem of Bianchi

The preceding reduction can be considerably generalized. Let Σ be any nth order system of ordinary differential equations:

$$(63) \qquad \frac{dx_i}{dt} = F_i(x_1, \cdots, x_n) \quad [i = 1, \cdots, n].$$

Suppose Σ is invariant under a group Γ of transformations $x \to \gamma(x)$ of $(x_1, \cdots, x_n)$-space, meaning by this that if $x(t)$ satisfies (63), then so do its transforms $\gamma(x(t))$, for all $\gamma \in \Gamma$. We shall show that this information is of real assistance in the explicit integration of (63).

In the case where Γ is a one-parameter group, the procedure is easy. In this case, except near singular points, Γ is locally equivalent[40] by a change of coordinates to the group of translations $y_1 \to y_1 + a$; $y_2, \cdots, y_n$ constant. Under these coordinates, (63) becomes $dy_i/dt = G_i(y_1, \cdots, y_n)$. Since subtraction of a constant from y_1 does not alter any dy_i/dt, we see that the G_i are really independent of y_1, giving

$$(64) \qquad \frac{dy_i}{dt} = G_i(y_2, \cdots, y_n) \quad [i = 1, \cdots, n].$$

Hence we reduce the integration of (63) to the integration of an $(n-1)$-st order system $dy_j/dt = G_j(y_2, \cdots, y_n)$ $[j = 2, \cdots, n]$, and the *quadrature* $y_1 = \int G_1(y_2(t), \cdots, y_n(t)) \, dt$.

More generally, let Γ be any r-parameter solvable Lie group of transformations of $(x_1, \cdots, x_n)$-space, which leaves (63) invariant. Then, almost by definition, Γ possesses a series of local[41] Lie subgroups $S_1 < S_2 < \cdots < S_r = \Gamma$, such that (i) S_{i-1} is normal in S_i, (ii) S_i is generated by S_{i-1} and a one-parameter subgroup Γ_i.

Locally, let us assume that the subsets of transitivity of S_{i-1} are the h-dimensional subspaces $y_{h+1}, \ldots, y_n$ constant, i.e. the parallels to the $(y_1, \ldots, y_h)$-plane for some h. Suppose further that by consideration of the invariance of (63) under S_{i-1}, we have been able to reduce the integration of (63) to the integration of

$$(65) \qquad \frac{dy_j}{dt} = G_j(y_{h+1}, \cdots, y_n) \quad [j = h+1, \cdots, n]$$

and quadratures. I shall show that the corresponding result holds for S_i.

There are two cases. If the subsets of transitivity of S_i are h-dimensional,

---

[40] See [63], p. 34. In general, I shall not give detailed references for the results from Lie theory assumed.

[41] The concept of a local Lie subgroup is explained in C. Chevalley, *Theory of Lie groups*, Princeton, 1946.

the result is trivial. Otherwise, since S_{i-1} is a normal (i.e. invariant) sub-group of S, the sets of transitivity [42] of S_{i-1} are permited nontrivially by Γ_i. By proper choice of coordinates, we can suppose that Γ_i effects the translations $y_{h+1} \rightarrow y_{h+1} + a$; $y_{h+2}, \cdots, y_n$ constant. Hence, as with (64), we can reduce the integration of (65) to the integration of

$$(65') \qquad \frac{dy_j}{dt} = G_j(y_{h+2}, \cdots, y_n) \quad [j = h+2, \cdots, n],$$

and the quadrature $y_{h+1} = \int G_{h+1}(y_{h+2}(t), \cdots, y_n(t)) \, dt$. This completes the proof by induction of the following

Theorem 2 (Bianchi [43]). *Let an n-th order system of ordinary differential equations Σ be invariant under a solvable Lie group having m-dimensional sets of transitivity. Then the integration of Σ can be reduced to the integration of an $(n-m)$-th order system, and quadratures.*

In §95, Γ_1 was the group $x \rightarrow x + a$, Γ_2 was $y \rightarrow y + b$, Γ_3 was $\theta \rightarrow \theta + \alpha$, $x \rightarrow x \cos \alpha - y \sin \alpha$, $y \rightarrow x \sin \alpha + y \cos \alpha$, and Γ_4 was $t \rightarrow t/\lambda$, $x \rightarrow x$, $v \rightarrow \lambda v$, etc.

97. Conclusion

Looking back over Chapters IV–V, we begin to realize the great importance of the group concept for fluid mechanics.

Thus, it underlies the entire theories of dimensional analysis and modeling. In the form of "inspectional analysis", it greatly generalizes these theories.

Again, the recognition of groups of symmetry often makes possible reductions in the number of independent variables involved in partial differential equations, directly through the method of search for symmetric solutions and the method of "separation of the time variable", and indirectly through "inverse methods". Moreover the method of search for symmetric solutions can be relied upon in general to give solutions locally (§89).

Even after the number of independent variables is reduced to one, so that no further reduction by the preceding methods is possible, the resulting system of ordinary differential equations can often be integrated most easily by use of group-theoretic considerations.

The preceding methods apply to non-analytic as well as to analytic, and to non-linear as well as to linear differential equations; thus they are free from the limitations of the usual methods of expansion in series or integrals. Hence group theory plays a fundamental role in solving the differential equations of fluid mechanics.

[42] By definition, a "set of transitivity" of S_{i-1} is, for some point y, the set Y of all $\sigma(y)[\sigma \in S_{i-1}]$. Since $\gamma \in \Gamma_i$ and $\sigma \in S_{i-1}$ implies $\gamma^{-1}\sigma\gamma \in S_{i-1}$, the set of all $\gamma(\sigma(y)) = (\sigma\gamma)(y)$ equals the set of all $\sigma(\gamma(y))$, and hence is also a set of transitivity of S_{i-1}.
[43] See [63] [36.5].

Finally, in Chapter VI, I shall try to show that group theory also underlies the classical force equations for a rigid body in a perfect (i.e. incompressible nonviscous) fluid.

I hope that, in the future, the debt of fluid mechanics to the concepts of group theory will be more explicitly recognized.

VI. *Added Mass*

98. Added mass of sphere

Qualitatively, the idea of added mass is a familiar one. For example, let a light paddle be dipped into still water and then suddenly given a rapid acceleration broadside. It is a matter of common experience that the apparent inertia (i.e. resistance to acceleration) of the paddle is greatly increased by the water around it. This increased inertia is what is called the "virtual mass" of the paddle, the difference between the real mass and the virtual mass being called the "induced mass" or "added mass".

Added mass was first given an exact mathematical interpretation by Green and Stokes, a little over a century ago.[1] The essence of their discussion went somewhat as follows.

Consider a sphere of mass m and radius a moving with speed v through an *incompressible nonviscous* fluid of density ρ (throughout the whole of the present chapter we shall consider only irrotational flows of such a "perfect fluid"). Without loss of generality, we may choose the axis of spherical coordinates as the direction of motion. Relative to the fluid at infinity, the velocity potential U is the dipole potential given in spherical coordinates by

(1)
$$U = \frac{-a^3 v \cos \theta}{2r^2}$$

In fact, one easily verifies that the normal derivative $\partial U/\partial r = v \cos \theta$ is the normal component of velocity of the surface of the sphere (§4). At a general point of the fluid, the radial and angular components of velocity are

$$u_r = \frac{a^3 v \cos \theta}{r^3}, \qquad u_\theta = \frac{1}{r}\frac{\partial U}{\partial \theta} = \frac{a^3 v \sin \theta}{2r^3}.$$

Hence the total kinetic energy of the fluid is

$$T = \int\int\int \tfrac{1}{2}\rho(u_r^2 + u_\theta^2) r^2 \sin \theta \, dr \, d\theta \, d\phi$$

$$= \pi \rho v^2 a^6 \int_{r=a}^{\infty} r^2 \, dr \int_0^{\pi} \sin \theta \left[\frac{\cos^2 \theta}{r^6} + \frac{\sin^2 \theta}{4r^6} \right] d\theta$$

[1] G. Green, *Mathematical Papers*, p. 315 (1833); [13], Vol. 1, p. 17, (1843). See [7], pp. 123–124 for a bibliographical discussion.

$$= \frac{\pi \rho v^2 a^6}{4} \int_{r=a}^{\infty} \frac{dr}{r^4} \int_0^{\pi} [1 + 3 \cos^2 \theta] \sin \theta \, d\theta$$

$$= \frac{\pi \rho v^2 a^6}{4} \left[\frac{-1}{3r^3} \right]_a^{\infty} \left[-\cos \theta - \cos^3 \theta \right]_0^{\pi}$$

$$= \frac{\pi \rho v^2 a^3}{3} = \frac{1}{2} \frac{2\pi \rho a^3}{3} v^2 = \frac{1}{2} m' v^2.$$

This gives the following classical result. The kinetic energy of the fluid is equal to that which would be possessed by a particle moving with the same speed as the sphere, and whose mass m' is equal to *half the mass of displaced fluid*.

Moreover it is evident that in a nonviscous fluid rotation of the sphere has no effect on the surrounding fluid; hence, the moment of inertia of the sphere should be unaffected. This suggests that (neglecting gravity) a sphere in such a fluid should be dynamically equivalent to a heavier sphere in vacuo whose "virtual mass" $m^* = m + m'$ is the mass m of the sphere plus an "added mass" m' equal to half the mass of displaced water, but whose moment of inertia is unchanged. This will be proved rigorously in §109, where it will be shown that the entire dynamic action of any irrotational, incompressible flow can be deduced from its kinetic energy, using the general equations of Lagrangian dynamics.

99. Applications

The preceding results have various simple applications. One is to the initial acceleration experienced by a spherical hydrogen balloon, suddenly released from a mooring. Suppose the mass of the balloon is 1/10 that of the displaced air. A person unaware of virtual mass would probably make the following erroneous computation. The total buoyancy force is, by Archimedes' Law, $9g$ times the mass of the balloon; hence (he would compute) the initial acceleration should be $9g$. In the case of a spherical hydrogen balloon immersed in water, an acceleration of at least $1,000g$ would be erroneously expected.

However, using the theory of virtual mass, it is easy to predict the correct initial acceleration. The *virtual mass* m^* of the balloon is $0.1 + 0.5 = 3/5$ that of the displaced air; hence the actual acceleration is $3g/2$. In water, it would be about $2g$.

A more subtle application arises if a weightless sphere is immersed in a liquid, and the whole liquid is suddenly given an acceleration a. What is the acceleration a^* of the sphere, relative to an outside observer? This problem may be treated as follows. The acceleration a is equivalent, for an observer in the fluid, to an apparent gravitational field of strength a.

Hence, by the reasoning just given, the initial acceleration $a^* - a$ of the sphere relative to an observer in the fluid should satisfy $a^* - a = 2a$, or $a^* = 3a$.

This prediction was confirmed by T. E. Caywood and the author,[2] for small air bubbles in water; it proved essential in interpreting experimental data on fluid flow patterns, like those of Plates I–II.

A third application is to pendulum clocks ([13], vol. 3, pp. 1–141). Added mass increases the inertia of a spherical bob by about 0.02% in air, retarding clocks having such bobs about 10 seconds per day, depending on the air density (pressure and temperature).

Many other applications could be cited (see §§103–4), but it seems more appropriate to discuss first the theoretical basis for calculating the added (or "induced") mass of a general nonspherical body.

As we shall see, this constitutes a beautiful topic in classical Lagrangian dynamics. It was created by Kelvin [82] and Kirchhoff [77], and is the main theme of Chapter VI of Lamb's *Hydrodynamics* [7].

100. Inertial Lagrangian systems

Consider the dynamical system consisting of a rigid body, Σ, and an ideal fluid without free surfaces bounded internally and/or externally by Σ. Clearly, Σ has six degrees of freedom of motion, which may be described by coordinates $q_1, \cdots, q_6$. Again, if $q(t)$ is given, then under very general conditions there is one and only one velocity potential (see §4 or [6], pp. 217, 311) $U = \dot{q}_i U^i(q)$ which tends to zero (is "regular") at infinity, satisfies $\nabla^2 U = 0$, and assumes on the surface S of Σ the values of $\partial U / \partial n$ given by the motion of Σ. Hence the fluid kinetic energy satisfies

$$(2) \qquad T = \tfrac{1}{2} \iiint \rho (\nabla U \nabla U)\, dR = \tfrac{1}{2} \sum_{i,j=1}^{6} T_{ij}(q) \dot{q}_i \dot{q}_j.$$

Moreover, the combined kinetic energy of fluid and solid satisfies a similar equation with different coefficients.

The symmetric matrix

$$(2') \qquad T_{ij}(q) = \rho \iiint (\nabla U^i \cdot \nabla U^j)\, dR$$

in (2) is called the "added mass" tensor; if the kinetic energy of Σ is added, the resulting matrix is called the "virtual mass" tensor.

The dynamical system just defined is non-holonomic, and has an infinite number of degrees of freedom if the fluid deformation is considered. However, it is natural to regard it as an ordinary Lagrangian system ([73], p. 36), having six degrees of freedom, and to think of the

---

[2] G. Birkhoff and T. E. Caywood, *J. Appl. Phys.*, **20** (1949), 646–59.

fluid configuration as defined by its boundaries moving under the "workless constraint" of incompressibility. In fact, this assumption is usually made without proof ([7], Ch. VI; [77], p. 238; and [82], p. 320.) We shall prove it in §109.

Moreover, by Avanzini's Theorem (§21, Theorem 1), the effect of gravity is simply to add a constant hydrostatic *buoyancy* to the system of inertial forces which would arise without gravity. Therefore, it suffices to consider the case $L = T$ of zero potential energy V, corresponding to $g = 0$. This defines a Lagrangian system in which the "generalized forces" Q_i satisfy

$$(3) \qquad Q_i = \frac{d}{dt}\left(\frac{\partial T}{\partial \dot{q}_i}\right) - \frac{\partial T}{\partial q_i}.$$

A Lagrangian system without potential energy may be called an *inertial Lagrangian system*; in §§ 101–112 we shall discuss the added mass (and virtual mass) coefficients defined by the inertial Lagrangian system (2)–(3).

101. Induced mass tensor

Near any reference position $q = 0$, it is convenient to let q_1, q_2, q_3 refer to translations of Σ parallel to the three coordinate axes, and to let q_4, q_5, q_6 refer to rotations (in radians) about these axes. This will define the $T_{hk}(0)$ in (2) as numbers, depending on the choice of axes in Σ.

For any such choice of axes, let U^1, U^2, U^3 be the velocity potentials induced by translations of Σ with unit velocity parallel to these axes; and let U^4, U^5, U^6 be the velocity potentials for rotation about these axes, with unit angular velocity. Then the fluid kinetic energy T in (2) satisfies

$$(4) \qquad 2T = \dot{q}_h \dot{q}_k \iiint \rho(\nabla U^h \cdot \nabla U^k)\, dR = \dot{q}_h \dot{q}_k M_{hk},$$

summed over repeated indices (the usual tensor convention). As in (2), $dR = dx_1 dx_2 dx_3$ is the differential of fluid volume; moreover, since $\nabla U^h \nabla U^k = \nabla U^k \nabla U^h$, evidently $T_{hk} = T_{kh}$; i.e. the induced (added) mass tensor is *symmetric*.

Under acceleration from rest, all $\dot{q}_h = 0$; hence (3) reduces to the simple form

$$(5) \qquad Q_h = T_{hk}(0)\ddot{q}_k \quad \text{if} \quad q = \dot{q} = 0.$$

This gives a simple meaning to T_{hk}. *It is the k-component of force, under unit acceleration from rest in the h-direction.* Moreover, since $T_{hk} = T_{kh}$, we obtain immediately the following Reciprocity Principle ([73], p. 305). The k-component of force under unit h-acceleration is equal to the h-component of force under unit k-acceleration.

In the simple case (5), we can easily verify directly that our system is Lagrangian. By Green's Second Identity ([6] p. 212), we have

$$(6) \qquad T_{hk} = \rho \int \int \int \nabla U^h \nabla U^k \, dR = \rho \int \int U^h \left(\frac{\partial U^k}{\partial n} \right) dS.$$

But now $\partial U^k / \partial n$ is equal (Chapter I, (7)) to the normal velocity component of Σ under unit velocity in the q_k direction. I shall introduce, for use now and later, the convenient notation $dS_k = (\partial U^k / \partial n) \, dS$, so that

$$dS_1 = dx_2 dx_3, \qquad dS_2 = dx_3 dx_1, \qquad dS_3 = dx_1 dx_2.$$
$$(7) \qquad dS_4 = x_2 dS_3 - x_3 dS_2, \qquad dS_5 = x_3 dS_1 - x_1 dS_3,$$
$$dS_6 = x_1 dS_2 - x_2 dS_1.$$

It is evident from the definition itself that

$$(8) \qquad T_{hk} = \rho \int \int U^h \, dS_k = \rho \int \int U^k \, dS_h.$$

It is also evident that if p denotes the scalar pressure, then $\int \int p \, dS_1$, $\int \int p \, dS_2$, $\int \int p \, dS_3$ are the components of *force* exerted by Σ on the fluid, while $\int \int p \, dS_4$, $\int \int p \, dS_5$, $\int \int p \, dS_6$ are the corresponding components of *moment*.

Now, consider the flow induced by unit acceleration from rest in the q_h-direction. It is easy to calculate that, initially, $U = 0$ and $\partial U / \partial t = U^h$; $U(x; t)$ differs from $tU^h(x)$ by infinitesimals of the second order in t. By Bernoulli's equation for pressure in an accelerated fluid (Chapter I, (5)),

$$(8^*) \qquad p + \tfrac{1}{2} \rho \nabla U \nabla U + \rho \frac{\partial U}{\partial t} = \text{const.},$$

since we have reduced to the case $g = 0$. This shows that the initial hydro-dynamic pressure p^h is everywhere equal to ρ times the "acceleration potential" U^h. Hence, substituting in (8), $T_{hk} = \int \int p^h \, dS_k$, and T_{hk} is *the k-component of force for unit acceleration from rest in the q_h-direction.* Specifically, Q_1, Q_2, Q_3 give the ordinary *force* components referred to the axes chosen, while Q_4, Q_5, Q_6 give the corresponding *moments*. This justifies the assumption (3), for the case (5) of acceleration from rest.

For motion under pure *translation*, the coordinates (q_1, q_2, q_3) can be used in the *large*. Hence the $T_{ij}(q) = T_{ij}(0)$ are *constant* and so (3) implies

$$(9) \qquad Q_i = \frac{d}{dt} \left(\frac{\partial T}{\partial \dot{q}_i} \right) - \frac{\partial T}{\partial q_i} = T_{ij} \ddot{q}_j \quad [i, j = 1, 2, 3].$$

This shows that the d'Alembert Paradox (§7) is implied by our assumption (3), which reminds us forcefully that our model does not correspond to physical reality in general.

The discussion of *moments* under translation, and of rotation effects generally, is more complicated (see §§ 111–112).

The preceding formulas refer to the induced or "added" mass. The total inertia or *virtual mass*, defined as the internal (rigid body) inertia of Σ plus the induced mass of the external fluid, is clearly another symmetric tensor (matrix), with just the same properties.

102. Special geometries; Rankine bodies

Theoretical coefficients of added mass have been calculated for various simple geometrical bodies besides spheres. They are commonly tabulated in dimensionless form in terms of the ratio k of the induced mass to the total mass ρ vol (Σ) of fluid displaced.

Many results, due to various authors, are derived in Lamb. For elliptic cylinders under translation and rotation, see [7], pp. 85, 88. For spheroids and ellipsoids, see [7], pp. 144, 146, 153, 159. For the case of sphere-pairs, see *ibid.*, p. 134.

The added mass of various other "two-dimensional" configurations (cylinders moving broadside) can also be calculated. Thus, J. L. Taylor[3] has calculated k for various polygons and parabolic digons. Circles and ellipses with symmetrically placed fins have also been treated by various authors,[4] with a view to analyzing the stabilizing effect of control surfaces on aircraft.

Among other special axially symmetric solids whose added mass has been determined analytically may be mentioned tori, spherical bowls, and "lenses" bounded by coaxial spherical caps.[5] Perturbations of the sphere have also been treated.

One can also treat *Rankine bodies*—solids of revolution whose effect in uniform potential flow parallel to the x_1-axis is the same as that of sources and sinks on this axis. We shall now treat such Rankine bodies, generalizing results of Max Munk and Sir Geoffrey Taylor.[6]

The first step of this analysis is to apply Green's Second Identity to

[3] J. L. Taylor, *Phil. Mag.*, 9 (1930), 161–83. For parallel plates, see D. Riabouchinsky, *Proc. Int. Math. Congress*, Strasbourg (1920), 568–85. See also W. G. Bickley, *Phil. Trans.*, A228 (1929), 235–74 and *Proc. Lond. Math. Soc.*, 37 (1934), 82–105, and B. N. Seth, *Publ. Lucknow Univ.* (1938–9).

[4] G. Kuerti et al., *Navord Rep. 2295* (1952); A. E. Bryson, *J. Aer. Sci.*, 20 (1953), 297–308 and 21 (1954), 424–26; R. C. Summers, ibid., 12 (1953), 856–57. Cf. formula (22).

[5] For tori, see W. M. Hicks, *Phil. Trans.*, 172 (1881), 609, and F. W. Dyson, *ibid.*, 184 (1892), 42. For spherical bowls, see A. B. Bassett, *Proc. Lond. Math. Soc.*, 16 (1885), 286. For lenses, see M. Shiffman and D. C. Spencer, *Quar. Appl. Math.*, 3 (1947), 270–88; L. E. Payne, *ibid.*, 10 (1952), 197–204. For nearly spherical solids, see G. Szego, *Duke Math. J.*, 16 (1949), 209–23; also Plana, *Mem. accad. sc. Torino*, 38 (1835), 209.

[6] *NACA Tech. Notes*, 104–6 and [80]. See also [7], pp. 164–66; W. Tollmien, *Ing.-Archiv*, 9 (1938); L. Landweber, *Quar. Appl. Math.*, 14 (1956), 51–56, and *J. Fluid Mech.*, 1 (1956), 319–36.

$U\nabla x_1 - x_1\nabla U$, recalling that $\nabla^2 U = \nabla^2 x_1 = 0$. Hence if S'' is a large sphere containing Σ, and R is the region between the surface S of Σ and S'', we get from (8), writing $U^h = U$.

(10)
$$T_{1h} = \rho \int\int_S U\left(\frac{\partial x_1}{\partial n}\right) dS$$

$$= \rho \int\int_S x_1 \frac{\partial U}{\partial n}\, dS - \rho \int\int_{S''} \left(x_1 \frac{\partial U}{\partial n} - U \frac{\partial x_1}{\partial n}\right) dS;$$

moreover on S'', $-\partial/\partial n = \partial/\partial r$. The integrals over S'' can easily be evaluated asymptotically, if we expand U in the form

(11) $$-U = \sum \frac{\mu_i x_i}{r^3} + O(r^{-3}), \qquad \frac{\partial U}{\partial r} = 2\sum \frac{\mu_i x_i}{r^4} + O(r^{-4}).$$

Since the area of S'' is $4\pi r^2$, the terms $O(r^{-3})$ resp. $O(r^{-4})$ contribute negligibly to (10). By symmetry, the terms in μ_2, μ_3 contribute nothing. To evaluate the remainder, we use spherical coordinates, writing $x_1 = r\sin\phi$, $dS = 2\pi d(\sin\phi)$. The integral over S'' differs by $O(r^{-1})$ from

$$2\pi\mu_1\rho \int_{-\pi/2}^{\pi/2} \{2\sin^2\phi + \sin^2\phi\}\, d(\sin\phi) = 2\pi\rho\mu_1 \left[\sin^3\phi\right]_1^1 = 4\pi\rho\mu_1.$$

Hence, passing to the limit as $r \to \infty$, we get

(12) $$T_{1h} = T_{h1} = 4\pi\rho\mu_1^h - \rho \int\int x_1 \left(\frac{\partial U^h}{\partial n}\right) dS,$$

where μ_1^h is the dipole moment of U^h in the x_1-direction. We note that the implication (10)—(12) is valid for *any* potential function U, regular at infinity, with $\int\int_{S''} (\partial U/\partial n)\, dS = 0$.

This generalizes a result of Sir Geoffrey Taylor [80], who considered the case $h = 1$. In this case $(\partial U^1/\partial n)\, dS$ is dS_1, and $\int\int x_1\, dS_1 = \int\int\int dR = \text{vol}(\Sigma)$, integrating by parts over the interior Σ of S. Hence,

(12') $$T_{11} = 4\pi\rho\mu_1^1 - \rho\,\text{vol}(\Sigma).$$

Taylor explained the presence of the term vol (Σ) in (12') by observing that a fluid-filled hollow in Σ increases its induced mass for translation by ρ times the volume of the hollow without changing the dipole moment at infinity of U^1.

He also noted that in the case of a *Rankine body*, $\mu_1^1 = \sum x_1^{(k)} e_k$. Hence, the added mass for translation along the x_1-axis is $4\pi\rho$ times the dipole moment of the source-and-sink distribution defining the body, minus the mass of fluid displaced. The formula for the sphere, with $\mu_1^1 = a^3/2$ as in §98, (1), is a special case:

$$T_{11} = 2\pi\rho a^3 - \rho \left(\frac{4\pi a^3}{3}\right) = \frac{2\pi\rho a^3}{3}.$$

The case of a prolate spheroid of revolution corresponds to a linear source distribution between the foci.

103. Theory and experiment

Though we have started with a theoretical discussion, the phenomenon of added mass was first discovered experimentally. In 1776, DuBuat[7] observed its effect on the period of small oscillations of spherical pendulum bobs. He obtained measured values of k in the range 0.45–0.67.

At the time, precise and consistent predictions of pendulum periods had a great scientific and technical interest. It was desired to know g and its anomalies accurately, and accurate chronometers were needed to determine longitude during long sea voyages. In a vacuum, one has $ml\ddot{\theta} = mg \sin \theta$, and hence g is related to the period τ of small oscillations of a pendulum of length l by the formula $g = 4\pi^2 l/\tau^2$. But corrections must be made for the effects of the air around the pendulum, as well as for friction of the suspension system.

Buoyancy will obviously reduce the restoring force $mg \sin \theta$ of a pendulum of density ρ by a factor $1 - (\rho'/\rho)$, where ρ' is the air density. Very careful measurements by Baily[8] had made it clear that this correction, amounting to about $5/\rho$ minutes a day, was insufficient. This lent considerable importance to the result of Poisson and Green that the "virtual mass" m^* should replace m in the left side of the pendulum equation, increasing τ further in the ratio $1 : [1 + (m'/m)]^{1/2}$. The *a priori* calculation of $m' = m^* - m$ for a spherical pendulum bob was especially striking.

Precise measurements focused attention on the fact that observed values of $m' = k\rho \text{ vol } (\Sigma)$, as inferred from the modified pendulum equation $(m + m')l\ddot{\theta} = m[1 - (\rho'/\rho)] \sin \theta$, systematically exceeded those predicted by the formulas of Poisson and Green.

The systematic discrepancy was explained by Stokes ([13], vol. 3, pp. 1–101) as an effect of *viscosity*. His analysis will be sketched in §115; it predicts an effect proportional to the Stokes' number $S = \nu^{1/2}/\omega^{1/2}a$, where ω is the frequency and a the sphere radius. He also predicted a viscous *damping*, again proportional to S in typical cases.

It follows that, for *rapid vibrations of small amplitude when S is small,*

[7] DuBuat, *Principes d'hydraulique*, 3d ed. Paris, 1816, 221–51 (first ed. in 1776). For a summary of early work, see ([13], vol. 3, pp. 76–122; or [75], pp. 97–101).

[8] F. Baily, *Trans. Roy. Soc.*, London (1832), 399–492.

ideal fluid theory should agree quite well with observation. In fact, fair agreement has been found in various experiments.[9] Unfortunately, the amplitudes have often not been recorded, and internal inconsistencies are often of the same magnitude as the viscosity corrections. Bessel's data seem peculiar; some of the extraneous effects involved were analyzed by Stokes ([13], vol. 3, p. 112).

Similarly, when all solid boundaries are remote, quite good agreement with ideal fluid theory is found for *acceleration from rest* during the first diameter or so of travel. However, after the sphere has travelled a few diameters, the flow separates (a vortex sheet is shed), and the steady state C_D becomes more important.[10] This occurs even sooner in the case of discs accelerated broadside;[11] this is to be expected, because the sharp edge promotes separation. In general, the tendency to separate depends on the total displacement in diameters; thus, in periodic motion it depends on $U_{\max}\,\tau/d$, where τ is the period.

104. Stability derivatives

Adaptations of the ideas suggested by the Lagrangian formulations of §§99–100 have been used to analyze the stability of steady motion for many kinds of solids moving through fluids. If the Lagrangian formulation were valid, then one could substitute (2) into (3) to get

$$(13) \qquad Q_i = \sum T_{ij}(q)\ddot{q}_j + F_i(q,\dot{q}),$$

the F_i being given by

$$(13') \qquad F_i = \sum \Gamma_i{}^{jk}(q)\dot{q}_j\dot{q}_k,$$

and $\Gamma_i{}^{jk} = (\partial T_{ij}/\partial q_k + \partial T_{ik}/\partial q_j - \partial T_{jk}/\partial q_i)/2$ being the Christoffel tensor ([73], p. 39). Moreover, the functions involved could, in principle, be determined *a priori* by potential theory, as claimed by Lagrange (§1).

In practice, only the *form* of equations (13)–(13') is assumed; (13') really corresponds to the assumption of "inertial scaling" (§70). The equations are usually applied only to small perturbations of steady motion, a restriction which permits one to treat the T_{ij} and $\Gamma_i{}^{jk}$ in (13)–(13')   as

[9] L. H. Laird, *Phys. Rev.*, 7 (1898), 102–5; S. Krishnayar, *Phil. Mag.*, 46 (1923), 1049–53; Y. T. Yu, *J. Appl. Phys.*, 13 (1942), 66–69, and 16 (1945), 724–29; T. H. Stelson, Ph.D. Thesis, Carnegie Inst. Tech. (1952). See also the refs. above, and those of §§104, 115.

[10] G. Cook, *Phil. Mag.*, 39 (1920), 350–82; P. Hirsch, *ZaMM*, 3 (1923), 93–107; G. Bagliarello, *Ric. Sci.*, 26 (1956), 437–61. For calculations of the distance to flow separation, see H. Tollmien, *Handbuch Exp. Phys.*, 4 (1931), Part 1, 272–79; S. Goldstein and L. Rosenhead, *Proc. Camb. Phil. Soc.*, 32 (1936), 392–401.

[11] J. Luneau, *C. R. Paris*, 227 (1948), 823–25, and 229 (1949), 227–28; H. W. Iverson and R. Balent, *J. Appl. Phys.*, 22 (1951), 324–28; five times the theoretical k is observed.

constants. These constants are usually defined *empirically* (like C_D, C_L, etc.), so as to fit observational data; the empirical constants defined in this way are then called *stability derivatives*.

Such stability derivatives were used during World War I, for example, to analyze the stability of spinning projectiles.[12] More modern applications are to the stability of rocket flight and of guided missiles; the case of fin-stabilized projectiles is much simpler.

An extremely important and highly developed application is to the stability of airplane flight. A closely related area of application is to the stability of airship (and submarine) motion. In these cases, the inertia of the air (water) surrounding the moving body is especially important.

Inspired by the speculations of Lagrange, various authors have tried to calculate stability derivatives from *a priori* considerations. Though some success has been had in the case of airships and submarines, considerable variations in real added mass have also been found experimentally under steady motion. The corresponding calculations of stability derivatives of airplanes are much more difficult: the calculations take into account circulation and vorticity distributions; a rather dubious use is made of the Joukowsky condition. For the details, we refer the reader to the technical literature.[13]

Technical interest in virtual mass also arises in connection with its effect on the natural frequencies of *ship* vibration, as well as of rolling, pitching, and heaving. In the first application, the effect of the free surface is easily estimated. Due to the relatively high frequency, one can assume that $U=0$ there. In the other applications, the effect of the surface waves generated is more complex. Again, we refer the reader to the literature.[14] Little systematic seems known about the dependence of coefficients on the Froude number.

Finally, the concept of virtual mass has been applied by various authors to the estimation of the impact forces encountered by seaplane floats in landing, and by missiles in entering the water and other liquids. For a brief summary of what is known, see ([15], pp. 243–50).

[12] Fowler, Gallop, Lock and Richmond, *Phil. Trans. Roy. Soc.*, *A221* (1920), 295–387, and *A222* (1922), 227–47. See E. J. McShane, J. L. Kelley, and F. V. Reno, *Exterior ballistics*, Denver, 1953.

[13] Reviews with bibliographies have been given by E. Reissner, *Bull. Am. Math. Soc.*, *55* (1949), 825–50 and I. E. Garrick, *Appl. Mech. Revs.*, 5 (1952), 89–91. See also W. R. Scruton, *Aer. Res. Comm. Rep.*, 1931. For airships, see Cowley and Levy, Frazer, Relf and Jones, *Adv. Comm. Aer.*, *Tech. Rep. 1918–19*, pp. 95–127; V. Szebehely, *Proc. Sec. Nat. Congr. Applied Mech. USA* (1954), 771–76.

[14] H. W. Nicholls, *Trans. Inst. Nav. Arch.* (1924), 141–63; F. M. Lewis, *Trans. Nav. Arch. Mar. Eng.*, 37 (1929), 1–18; E. B. Moullin, *Proc. Camb. Phil. Soc.*, *24* (1928), 400–13 and 531–58; A. D. Brown et al., *ibid.*, *26* (1930), 258–62; G. Weinblum, *Schiffbau*, *32* (1931), 488–95, 509–11 and 525–29; M. D. Haskind, *Izv. Akad. Nauk* (1946), 23–34, and *Prikl. Mat. Meh.*, *10* (1946), 475–80; K. Wendel, *Jahr. der Schiffsbau Ges.*, *44* (1950), 207–55.

Although deductive theories of §§99–102 are not rigorously applicable to any of the above examples, the concept of an "added mass" tensor has proved fruitful in all of them.

105. Added mass and momentum

In most applications, virtual mass effects are combined with effects due to many other causes, the discussion of which is interesting only to specialists. We therefore return now to the pure theory of virtual mass, which is very appealing esthetically and mathematically. Before going on to the more abstract aspects of the subject in §§108–112, we shall present a few specific results which should add meaning to the abstractions involved.

The most satisfactory definition of induced mass coefficients is through kinetic *energy* integrals, as in (2) and (4). These integrals *converge* at infinity, because (§7) $\nabla U = O(r^{-3})$ in space. The kinetic energy integral also converges at infinity in the case of plane Dirichlet flows; in this case $\iint (\nabla U \nabla U)\, dx dy = O(\int r^{-4} r\, dr)$ is finite.

We shall now interpret the T_{hk} as momentum integrals. It has been noted by various authors[15] that the momentum integrals diverge in the usual sense. Hence one must take care when interpreting the T_{hk} in terms of momentum. Now for the details:

The T_{hk} of (8) are integrals over the boundary of S of Σ, which can be written in a new notation as

$$(14) \qquad T_{hk} = \rho \iint_S U\, dS_K,$$

$U = U^h$ being harmonic and $dS_K = \sum_{i=1}^3 K_i dS_i$ expressing the differential of flux of a vector field $K = (K_1(x), K_2(x), K_3(x))$ through S. Thus in the case $k = 1$ of translations parallel to the x_1-axis, $K = (1, 0, 0)$; in the case of rotations about the x_1-axis, $K = (0, x_3, -x_2)$, and so on. This convenient notation defines a useful class of Stieltjes integrals over surfaces in space whenever $\iint |dS_K|$ is finite. Note that whenever K is the velocity-field of a rigid motion, div $K = 0$. This condition and the condition that U is a harmonic function "regular" at infinity ([6], p. 217) are all that we require for the argument below.

We define a "K-line" as an integral curve of the system $dx_i/dt = K_i(x)$, or $dx/dt = K$ $(dx_i = K_i dt)$. Thus if K represents translation parallel to the x_1-axis, the K-lines are the parallels to this axis; if K corresponds to screw motion about the x_1-axis, the K-lines are helices about this axis, etc. We then apply the Divergence Theorem to the vector field UK, over

[15] [7], p. 161; H. Tollmien, *ZaMM*, *18* (1938), p. 154.

regions R bounded by surfaces $S \cup S' \cup S''$, where S'' consists of K-lines. Since $dS_K \equiv 0$ on the surface S' consisting of K-lines, and since

$$\text{div}\,(UK) = K_i \frac{\partial U}{\partial x_i} + U\,\text{div}\,K,$$

we get from (14), using div $K = 0$,

$$(14') \qquad T_{hk} = \rho \iiint_R \left(K_i \frac{\partial U}{\partial x_i}\right) dR - \rho \iint_{S'} U\,dS_K.$$

Here the relation between the signs of the double and triple integrals involved is correct, provided surface integrals are taken *into* R. In particular, if R is an *infinite* region, if S' is removed to infinity, and if the integral over R converges, we get

$$(14^*) \qquad T_{hk} = \rho \iiint_R \left(K_i \frac{\partial U}{\partial x_i}\right) dR.$$

We shall now consider various cases corresponding to special induced mass coefficients.

If $k=1$, then $K=(1, 0, 0)=\text{grad}\,x_1 = \nabla x_1$ in various notations. We let S'' be an infinitely long *cylinder* parallel to the x_1-axis and containing Σ. Then since the momentum integral

$$(15a) \qquad T_{h1} = T_{1h} = \rho \iint_S U^h\,dx_2 dx_3 = \rho \iiint_R \left(\frac{\partial U^h}{\partial x}\right) dR$$

converges at infinity, we get the following result. *The induced mass coefficients corresponding to translation along the x_1-axis are equal to the x_1-momentum inside any infinite cylinder parallel to the x_1-axis and containing Σ, under motion with unit h-velocity.* It is simplest to take R as the cylinder circumscribed about Σ. This result, as regards T_{11}, was obtained by Theodorsen [81].

Again, if $k=4$, then $K=(0, x_3, -x_2)=x_3\nabla x_2 - x_2\nabla x_3$. We let S'' bound any *solid of revolution* containing Σ and having the x_1-axis as axis of symmetry. Then the boundary $S \cup S''$ of R consists of S and K-lines (circles), so that (14*) reduces to

$$(15b) \qquad \begin{aligned} T_{h4} &= \rho \iiint_R \left(x_2 \frac{\partial U^h}{\partial x_3} - x_3 \frac{\partial U^h}{\partial x_2}\right) dR \\ &= \rho \iiint_R \left(\frac{\partial U^h}{\partial \theta}\right) dR, \end{aligned}$$

where $\theta = \tan^{-1} x_3/x_2$. Hence T_{h4} is the *moment of momentum* of the fluid in R, about the x_1-axis.

The other T_{hk} are easily obtained from (15a)–(15b) by cyclic permutations of axes. Also, the case of *screw motion* about the x_1-axis, $K = \alpha \nabla x_1 + \beta(x_2 \nabla x_3 - x_3 \nabla x_2)$ is easily obtained by superposition of the two preceding formulas. We let S'' be a *circular cylinder* circumscribed about Σ. Hence the induced inertia is the screw momentum $\rho \iint (\alpha \partial U^h / \partial x_1 + \beta \partial U^h / \partial \theta) \, dR$ of the fluid inside S''; this is the same for all circular cylinders parallel to the x_1-axis which contain Σ.

106. Other interpretations

Quite recently, Sir Charles Darwin[16] has given a fresh and very simple interpretation of the (ideal) added mass of a rigid body Σ, in terms of the *drift*, or net displacement of a transverse fluid surface induced by the translation of Σ from $-\infty$ to $+\infty$ along a given axis. He has shown that the *induced volume* μ_1/ρ, defined as the added mass under translation divided by the fluid density, is equal to the volume enclosed between the initial and final positions of any such surface.

For *plane* flows about a plane area Σ, added mass has another fairly simple interpretation, in terms of the conformal transformation

$$(16) \qquad z' = az + c_0 + \frac{c_1}{z} + \frac{c_2}{z^2} + \cdots, \qquad a > 0;$$

mapping the exterior of the unit circle onto the exterior of Σ. As already remarked in §8, there is one and only one such transformation. Observe that, if V is the stream function, the differential $(\nabla V \cdot \nabla V) dx \, dy$ of kinetic energy is conserved under conformal transformation, and that the velocity at infinity is changed in the ratio $1{:}a$. On the other hand, $V + u_\infty y = \psi$ satisfies the boundary condition $V = 0$ on Σ.

A straightforward asymptotic calculation using these facts ([79], p. 204) shows that the added mass M_{11} of Σ satisfies

$$(17) \qquad M_{11}/\rho + \text{Area} \ (\Sigma) = 2\pi a^2 [1 - \text{Re} \ (c_1)],$$

for flows parallel to the x-axis. Formula (17) reduces the calculation of added mass to a problem in conformal transformation.

Another interesting conclusion is Polya's theorem, that a circular disc has the smallest average added mass (averaged over all possible orientations) of all plane domains with given area. The corresponding result for spheres in space has recently been proved by Schiffer.[17]

Finally, there is a beautiful result relating added mass to the theory

---

[16] *Proc. Camb. Phil. Soc.*, *49* (1953), 342–54. See also M. J. Lighthill, *J. Fluid Mech.*, *1* (1956), 31–53, and 311–12.

[17] G. Polya, *Proc. Nat. Acad. Sci.*, *33* (1947), 218–21; M. Schiffer, *Comptes Rendus Paris*, *244* (1957), 3118–21.

of free boundaries, treated in Chapter III. As first proved by Riabouchin-sky, *free boundaries extremalize added mass* with respect to all boundary variations enclosing the same volume (or area, in plane flows). For the derivation and applications of this theorem, we refer the reader to ([15], pp. 85–89 and 177–84).

107. Local normal form

As already explained in §101, the virtual mass tensor of a body Σ refers to a particular choice of axes in the body, and applies to perturbations of its position about a given reference position $q=0$. The sphere is a very simple case. If axes are taken through the center of the sphere, then, in the notation of §100, $T_{11}=T_{22}=T_{33}=m^*$, $T_{44}=T_{55}=T_{66}=2ma^2/5$, and all T_{ij} $[i\neq j]$ off the main diagonal are zero.

Another familiar special case concerns a rigid body *in vacuo*. If the origin is taken at the center of gravity, then all T_{ij} $[i\neq j$ and i or $j=1, 2, 3]$ vanish, while $T_{11}=T_{22}=T_{33}=m$, the mass of the body. Again, by choosing Cartesian axes as the principal axes of inertia, we can make all T_{ij} $[i\neq j$ and $i,j=4, 5, 6]$ vanish. Hence we can characterize the inertia tensor by four scalars T_{11}, T_{44}, T_{55}, T_{66}, which can be reduced to *two* by a change of length and time units. Further, $T_{44}+T_{55}\geq T_{66}$ and cyclically; the case of an ellipsoid is perfectly general.

In the preceding two cases, the matrix $\|T_{ij}\|$ was reduced to diagonal form by proper choice of axes. It is interesting to see how far one can simplify the matrix of inertial coefficients T_{ij}, by appropriate choice of Cartesian axes and a "center of virtual inertia", in the general case of a fluid of positive density. This is a simple exercise in the theory of quad-ratic forms.

The example of the paddle (§98) shows that the "virtual inertia" for translation can be different in different directions; sidewise acceleration is much easier to produce than broadside acceleration. However, since any quadratic form is equivalent[18] to a diagonal form under the rotation group, we can always rotate axes so as to get $T_{12}=T_{23}=T_{31}=0$, and "virtual inertias" of T_{11}, T_{22}, T_{33} in "principal directions" of trans-lation.

Except in the degenerate case of a weightless plane lamina, where one of the T_{ii} $[i=1, 2, 3]$ vanishes, we can achieve a further simplification by proper choice of the origin as a "virtual center of inertia." Let w_1, w_2, w_3 denote rotation at one radian per second about one set of axes parallel

---

[18] The algebraic theorems assumed here are proved for example in [42], Ch. IX, Thm. 21. For the results see [7], §126; the general case was treated by Clebsch, *Math. Annalen*, *3* (1870), 238. For related results, see G. W. Morgan, *Quar. Appl. Math.*, *12* (1954), 277–85.

to the principal directions of translation; let X, Y, Z denote the translations at unit speed in the principal directions; and let w_1', w_2', w_3' denote rotations about axes translated through the vector (x, y, z). Then

(18)
$$w_1' = w_1 + yZ - zY,$$
$$w_2' = w_2 + zX - xZ,$$
$$w_3' = w_3 + xX - yX.$$

The substitution of w_1' for w_1 does not change the interaction energy T_{14} between X and w_1, since there is none between X and Y or Z. On the other hand, it increases $T_{15} = T_{51}$ by zT_{11}, increases $T_{16} = T_{61}$ by xT_{22}, decreases T_{24} by zT_{22}, and so on.

Hence we can make $T_{15} = T_{24}$ by proper choice of z, and similarly $T_{16} = T_{34}$ and $T_{26} = T_{35}$. Thus *we can reduce the matrix of inertial coefficients* to the simplified form of Fig. 27.

Translation			Rotation		
α	0	0	ρ	ξ	η
0	β	0	ξ	σ	ζ
0	0	γ	η	ζ	τ
ρ	ξ	η	T_{44}	T_{45}'	T_{46}'
ξ	σ	ζ	T_{45}	T_{55}	T_{56}
η	ζ	τ	T_{46}	T_{56}	T_{66}

FIG. 27.

In this canonical form we have *fifteen arbitrary constants*, which can be reduced to thirteen by a change of length and time unit. That is, the general case involves *thirteen* dimensionless ratios and two choices of unit.

If a body has three perpendicular planes of symmetry, like an ellipsoid, we can take these as coordinate planes. Reflection in (say) the (x, y)-plane leaves Z, w_1, w_2 fixed but reverses the signs of X, Y, w_3; it also leaves kinetic energy unchanged. Hence $T_{31} = -T_{31} = 0$, and similarly

$$T_{32} = T_{36} = T_{41} = T_{42} = T_{46} = T_{51} = T_{52} = T_{56} = 0.$$

Repeating the argument for the other coordinate planes, we see that the matrix of inertial coefficients can be diagonalized. Thus we have *six* arbitrary constants and *four* dimensionless ratios. (There are two in the case of a rigid body in vacuo.)

Another interesting case is that of symmetry in three perpendicular axes, though not in the planes through them. This is typified by the helix $x = r\cos z$, $y = r\sin z$, $|r| \leq 1$, $|z| \leq 2\pi$. Symmetry in the z-axis leaves Z, w_3 fixed but reverses the signs of X, Y, w_1, w_2. Hence as above,

$$T_{13} = T_{23} = T_{43} = T_{53} = T_{16} = T_{26} = T_{46} = T_{56} = 0.$$

Repeating the argument for the other coordinate planes, we see that all inertial coefficients off the principal diagonal vanish except the "helical products of inertia," ρ, σ, τ. We thus have *nine* coefficients of inertia, which we can reduce after changes of length and time scales to *seven* essentially independent parameters.

The preceding arguments can be applied equally well to the added mass tensor—though the principal axes will be different in general, unless given by symmetries. I treated the virtual mass tensor to include the familiar case of a rigid body *in vacuo* as a special case.[19]

108. Geometrical aspects

We shall now treat ideal virtual mass as a branch of pure geometry. We begin with the result of §§100–1, that the system consisting of a rigid body Σ in an ideal fluid is an *inertial Lagrangian system* with kinetic energy $T = \frac{1}{2} T_{hk} \dot{q}_h \dot{q}_k$. From this, one easily deduces the classical[20] result that the "natural" trajectories, which would arise in the absence of external forces, are *geodesics*. Specifically, $Q = 0$ in (3) if and only if $\int T \, dt$ is a minimum. This obvious application of Euler's equations is the simplest case of the Principle of Least Action, the variational formulation of dynamics.

More precisely, the space of "configurations" $q = (q_1, \cdots, q_6)$ of the system is a Riemannian variety relative to the "arc-length"

$$(19) \qquad ds^2 = \sum T_{ij}(q) dq_i dq_j.$$

Moreover, by energy conservation, ds/dt is constant, and so $\int T \, ds$ as well as $\int T \, dt$ is an extremum (and local minimum). These principles are easy to verify in familiar examples.

Thus if $V = V_2$ is a frictionless surface $x = x(q_1, q_2)$ in ordinary space, we see that the force of constraint is perpendicular to V_2; hence in the absence of external forces the normal to the particle path is normal to V_2; this condition is well known to characterize geodesics. More generally, consider an arbitrary trajectory γ on V_2. The force-component of constraint, normal to V_2, is usually neglected. The remaining force is in the plane tangent to V_2. It has two components: the component $\ddot{s}$ *tangent* to γ, which can be computed using $T = \frac{1}{2} \sum \dot{x}_k{}^2$ as $\ddot{s} = Q_1 \dot{q}_1 + Q_2 \dot{q}_2$, and the component *normal* to γ in the plane tangent to V_2, which is $v^2 = \dot{s}^2$ times the geodesic curvature.

Corresponding formulas hold in any Riemann space V. In particular, Q transforms like a (contravariant) *vector*, and its normal component is

---

[19] For the material of §107, see [32].
[20] H. Hertz, *Principles of mechanics*, §357; [73], §100; J. L. Synge, *Phil. Trans.*, A226 (1926), 31–106.

v^2 times the vector geodesic curvature. Hence the *dynamical* problems of inertial Lagrangian systems are equivalent to *geometrical* problems.

109. Proof that system is Lagrangian

I shall now prove, for the case of a finite body in space, the validity of the assumption that the generalized forced Q_i, defined by the variational equations (3) of Lagrange, are in fact the components Q_i^* of resultant pressure thrust resp. moment in the ordinary sense.[21] These are, of course, defined mathematically as integrals over the body surface

$$(20) \qquad\qquad Q_i^* = \iint p N_i \, dS.$$

Here N_i denotes the normal component of surface displacement under translation resp. rotation, corresponding to the i-th generalized coordinate, and p is defined by the Bernoulli equation

$$(21) \qquad\qquad p + \rho\left[\tfrac{1}{2}\nabla U \nabla U + \frac{\partial U}{\partial t}\right] = p_0(t),$$

for accelerated motion in an ideal fluid (Chapter I, (5)), provided that hydrostatic buoyancy forces are neglected as usual.

A rigorous proof is made difficult by the facts that the total mass involved is infinite, and that the "configuration space" of the fluid has an infinite number of dimensions—corresponding to the number of degrees of freedom of fluid motion. The only discussions of these difficulties in the literature are due to Lamb, ([78], and [7], Sections 135–136) and seem not fully satisfactory.[22] I shall therefore present a new and very elegant variational proof due, with minor modifications, to Mr. John Breakwell. In presenting this proof, I shall follow the suggestive δ notation for variational differentials generally used in dynamics, even though the derivative notation is preferred by most modern authors writing about the calculus of variations.[23]

Two general theorems will be used, without proof. The first is simply Euler's identity for the first variation

$$(22) \qquad\qquad \delta \int_{t_0}^{t_1} T \, dt = \int_{t_0}^{t_1} Q_i \delta q_i(t) \, dt,$$

---

[21] The validity of assuming (3) was questioned by Boltzmann (*Borchardt's Jour.*, vol. *73*, p. 111) and Purser. See also R. von Mises, *ZaMM*, *4* (1924), 155–81 and 191–213.

[22] Thus in [78], the integrals involved are not convergent, and the exact assumptions underlying the discussion of [7] seem somewhat obscure.

[23] M. Morse, *The Calculus of Variations in the Large*, New York, 1934; G. A. Bliss, *Lectures on the Calculus of Variations*, Chicago, 1946.

with the Q_i as in (3), and summation with respect to repeated indices being understood. The second is the fact that, in (22), all variations $\delta q_i(t)$ are possible, subject only to $\delta q_i(t_0) = \delta q_i(t_1) = 0$. This amounts to saying that the configuration space is "holonomic" for the rigid body. It follows that, in order to prove the identity $Q_i = Q_i^*$ in comparing (3) and (2), it is sufficient to prove the first identity of

$$(23) \qquad \int_{t_0}^{t_1} Q_i^* \delta q_i(t) \, dt \overset{?}{=} \delta \int_{t_0}^{t_1} T \, dt = \int_{t_0}^{t_1} dt \Big\{ \delta \iiint \tfrac{1}{2} \nabla U \nabla U \, dm \Big\}.$$

The second identity is obvious, writing dm for the mass-differential $\rho \, dR$ in (23).

One can prove (23) by manipulating its right-hand term as follows. The manipulation is permissible since $\nabla U \nabla U = O(1/r^6)$ as before, whence the quadruple integrals over space-time are absolutely convergent, and interchange of the order of integration is allowed. First, using "Lagrangian" coordinates moving with the fluid, we get immediately

$$(24) \qquad \begin{aligned} \int_{t_0}^{t_1} dt \Big\{ \delta \iiint \tfrac{1}{2} \nabla U \nabla U \, dm \Big\} &= \int_{t_0}^{t_1} dt \Big\{ \iiint u_i \delta u_i \, dm \Big\} \\ &= \iiint dm \Big\{ \int_{t_0}^{t_1} u_i \delta u_i \, dt \Big\}, \end{aligned}$$

where $u_i = \dfrac{\partial U}{\partial x_i}$. Integrating by parts, we get for each fluid particle

$$(25) \qquad \int_{t_0}^{t_1} u_i \delta u_i \, dt = \Big[u_i \delta x_i \Big]_{t_0}^{t_1} - \int_{t_0}^{t_1} a_i \delta x_i \, dt,$$

where $a_i = du_i/dt$ denotes acceleration. By the equations of motion (Chapter I, (2)), with the buoyancy force of gravity neglected (§21, Theorem 1), $-a_i = \partial p/\rho \partial x_i$, hence

$$(26) \qquad \begin{aligned} \operatorname{div}(p \delta x) &= \sum \delta x_i \Big(\frac{\partial p}{\partial x_i} \Big) + p \operatorname{div}(\delta x) \\ &= \sum \delta x_i \Big(\frac{\partial p}{\partial x_i} \Big) = - \rho \Sigma a_i \delta x_i, \end{aligned}$$

since $\operatorname{div}(\delta x) = 0$ in an incompressible fluid. Similarly, $\operatorname{div}(U \delta x) = \sum u_i \delta x_i$. Substituting back from (25) in the last expression of (24), and from (26) and the other divergence formula in (24), we get

$$(27) \qquad \iiint dR \Big[\operatorname{div}(\rho U \delta x) \Big]_{t_0}^{t_1} + \iiint dR \Big\{ \int_{t_0}^{t_1} \operatorname{div}(p \delta x) \, dt \Big\}.$$

Since $p = O(1/r^6)$, $\delta x = O(r)$, and $dR = 4\pi r^2 dr$, the quadruple integral is absolutely convergent as before. Hence we can reverse the order of integration, and then apply the Divergence Theorem[24] to get

$$(28) \qquad \delta \int_{t_0}^{t_1} T \, dt = \left[\int \int_S \rho U \delta x_n \, dS \right]_{t_0}^{t_1} + \int_{t_0}^{t_1} dt \left\{ \int \int_S p \delta x_n \, dS \right\}.$$

Here δx_n denotes the component of δx normal to the common boundary S of the solid and liquid. Since the solid and liquid are in contact, $\delta x_n = \sum N_i \delta q_i$, in the notation of (20). In particular, $\delta x_n = 0$ at t_0 and t_1, and the first term on the right hand side of (28) vanishes, since $\delta q_i(t_0) = \delta q_i(t_1) = 0$. From $\delta x_n = \sum N_i \delta q_i$, we get also

$$\int \int_S p \delta x_n \, dS = \sum \int \int p N_i \delta q_i \, dS = \sum Q_i^* \delta q_i,$$

by (20). Substituting back from these results into (28), we get

$$(29) \qquad \delta \int_{t_0}^{t_1} T \, dt = 0 + \int_{t_0}^{t_1} Q_i^* \delta q_i(t) \, dt.$$

This proves (23), and hence $Q_i^* = Q_i$.

110. Homogeneity

The Riemannian variety V defined by (19) from the configuration space of a rigid body Σ in an infinite ideal fluid is remarkable, in that it possesses a (simply) transitive group of "isometries" (rigid motions), leaving ds invariant. In modern mathematical terminology, it is a *homogeneous space*. This is because of an obvious group-theoretic Relativity Principle: *Relative to the body, all positions seem equivalent.* We can express this formally as follows.

The different positions $q = a, b, c, \ldots$ of the body in space correspond one-one to the different rigid motions $\alpha, \beta, \gamma, \ldots$ carrying the body from a fixed reference position 0 to $a, b, c, \ldots$. Hence we can identify the points of the configuration-space with the elements of the *Euclidean group* [42, p. 259]. Moreover if α is any particular rigid motion, the position $\alpha\sigma$ appears just the same to an observer in position a, as σ does to an observer at 0, all Cartesian coordinate systems being equivalent. Hence the "group-translation" $\sigma \to \alpha\sigma$ cannot affect the kinetic energy metric (19).

Now let α vary; regarding V as the group manifold of the Euclidean

[24] To justify this, we need the theorem, due to Lagrange, that the fluid particles in contact with the rigid body form an invariant set ([11], Vol. 1, p. 97). Also, we need to remark that the surface integrals, over large spheres, of $U\delta x_n$ and $p\delta x_n$, tend to zero, since $\delta x_n = O(\int u \, dt) = O(1/r^3)$.

group, we see that V has a "simply transitive"[25] group of "isometries" (i.e. of motions leaving the metric ds^2 invariant). I shall call such a manifold, a *Riemannian group manifold*; we can always regard the isometries as left-translations.

Moreover another group-theoretic remark is possible. By a "steady motion" is meant in dynamics a motion which is independent of time relative to body axes. As in (13), an acceleration $\ddot{q}$ relative to steady motion increases $Q_i = \sum d(T_{ij}\dot{q}_j)/dt + (\partial T_{hk}/\partial q_i)\dot{q}_h\dot{q}_k$ by $T_{ij}\ddot{q}_j$. Hence to obtain the forces for an arbitrary motion, we can simply superpose those for an acceleration $\ddot{q}$ from rest, discussed in §§100–2, onto those for steady motion. Thus if we can determine the forces on a rigid body under *steady motion* in an "ideal" (i.e. incompressible nonviscous) fluid, we can determine the forces under any motion. We therefore concentrate on the problem of finding the forces in steady motion.

The only kinds of steady motion which are geometrically possible for a rigid body in Euclidean space are well known[26] to be translation, rotation, and screw motion with a fixed pitch—all at constant velocity.

By definition, if $\alpha(t)$ is a steady motion, the displacement σ required to get from $\alpha(t)$ to $\alpha(t+h)$ is a function $\sigma(h)$ of h alone. Hence

$$\alpha(0)\sigma(h+h') = \alpha(h+h') = \alpha(h)\sigma(h') = \alpha(0)\sigma(h)\sigma(h').$$

Left-cancelling $\alpha(0)$, we have $\sigma(h+h') = \sigma(h)\sigma(h')$; hence $\sigma(h)$ is a *one-parameter subgroup* under canonical parameters—and $\alpha(t)$ is an image thereof under an isometry, namely, the group translation $\sigma(t) \to \alpha(0)\sigma(t) = \alpha(t)$.

Moreover since the total kinetic energy is constant in steady motion, clearly $\dot{s} = v$ is constant in the associated Riemannian variety V. Hence by §108, under steady motion, the force vector Q is a constant v^2 times the vector geodesic curvature of the one-parameter subgroup $\sigma(h)$. Hence *the force on a rigid body under steady motion in an ideal fluid is proportional to the vector geodesic curvature of the corresponding one-parameter subgroup of the Euclidean group V, for a suitable "left-invariant" metric of V.* Moreover this left-invariant metric is determined at all points by the "inertial coefficients" $T_{ij}(0)$ already discussed in §§100–2.

111. Concepts from Lie theory

I shall now derive a formula for the geodesic curvature of the one-parameter subgroups of an arbitrary Riemannian group manifold. This

---

[25] By this it is meant that, given σ, $\tau \in V$, there exists one and only one α with $\sigma\alpha = \tau$. We are here assuming some slight acquaintance with the left-translations of an abstract group.

[26] See for example J. S. Ames and F. D. Murnaghan, *Theoretical Mechanics*, p. 87.

result is incidentally of interest for the geometry of Lie groups—which illustrates once again the essential unity of all mathematics.

It would be impossible for me to give enough of the theory of Lie groups, in this short space, to make clear all details of the derivation. However, I should like to give enough to make clear the meaning of the final formula, at least in the case of the Euclidean group.

If a rigid body is moving with unit velocity parallel to the x_1-axis, the velocity of any molecule of the body is $(1, 0, 0)$. Hence if $F(x_1, x_2, x_3)$ is any function defined over space, the rate of change of the value of the function, as observed from the molecule, is $\partial F/\partial x_1$. The operator $\partial/\partial x_1$ defined in this way is called the Lagrangian symbol expressing the *infinitesimal transformation* associated with the rigid motion of translation parallel to the x_1-axis.

If the rigid body were rotating with unit angular velocity (one radian per second) around the x_1-axis, a molecule at (x_1, x_2, x_3) would have velocity $(0, -x_3, x_2)$. Hence the rate of change in $F(x_1, x_2, x_3)$, as observed from the molecule, would be $x_2 \partial F/\partial x_3 - x_3 \partial F/\partial x_2$. The infinitesimal transformation associated with rotation around the x_1-axis is therefore expressed by the Lagrangian symbol (linear differential operator) $x_2 \partial/\partial x_3 - x_3 \partial/\partial x_2$.

Associated with the six degrees of freedom of motion of a rigid body, we therefore have the six infinitesimal transformations expressed by

$$(30) \qquad
\begin{aligned}
E_1 &= \frac{\partial}{\partial x_1}, & E_4 &= x_2 \frac{\partial}{\partial x_3} - x_3 \frac{\partial}{\partial x_2}, \\[2ex]
E_2 &= \frac{\partial}{\partial x_2}, & E_5 &= x_3 \frac{\partial}{\partial x_1} - x_1 \frac{\partial}{\partial x_3}, \\[2ex]
E_3 &= \frac{\partial}{\partial x_3}, & E_6 &= x_1 \frac{\partial}{\partial x_2} - x_2 \frac{\partial}{\partial x_1}.
\end{aligned}$$

Or, by the corresponding vector fields (velocity fields) $(1, 0, 0)$, $(0, 1, 0)$, $(0, 0, 1)$, $(0, -x_3, x_2)$, $(x_3, 0, -x_1)$, $(-x_2, x_1, 0)$.

The effect of the velocity-field (infinitesimal transformation) E_i, operating for a time t, is denoted $\exp(tE_i)$; thus $\exp(2E_4)$ would denote rotation about the x_1-axis through two radians. If $t < 0$, $\exp(tE_i)$ is taken to denote the inverse of the transformation $\exp(-tE_i)$. Hence we have, for all real t, u, the identity,

$$(31) \qquad \exp(tE_i)\exp(uE_i) = \exp(\{t+u\}E_i).$$

By *canonical parameters* for (say) the Euclidean group, we mean a parametric representation of rigid motions by vectors, such that the rigid motion associated with the vector $t = (t_1, \cdots, t_6)$, is the finite transformation

$$(32) \qquad \exp(t_1 E_1 + \cdots + t_6 E_6),$$

which expresses the total displacement occurring when the velocity-field $t_1 E_1 + \ldots + t_6 E_6$ operates for a unit time.

Finally, the *Poisson bracket* or *commutator* $[E_i, E_j]$ of two infinitesimal transformations E_i and E_j is defined as the double limit

$$(33) \qquad \lim_{t,u \to 0} \left[\frac{1}{tu} \exp(-tE_i) \exp(-uE_j) \exp(tE_i) \exp(uE_j) \right].$$

It is well known that this limit is the differential operator[27]

$$(34) \qquad E_i E_j - E_j E_i = \sum_{h,k} \left(E_j^{(h)} \frac{\partial E_i^{(k)}}{\partial x_h} - E_i^{(h)} \frac{\partial E_j^{(k)}}{\partial x_h} \right) \frac{\partial}{\partial x_k},$$

which can be easily computed. Thus with the Euclidean group, we have for example

$$(35) \qquad \begin{aligned} [E_1, E_2] &= [E_1, E_4] = 0; \\ [E_1, E_5] &= -E_3; \\ [E_4, E_5] &= -E_6. \end{aligned}$$

From these identities, all other Poisson brackets of $E_1, \cdots, E_6$ can also be computed by cyclic permutations of the subscripts and use of the obvious identity $[E_i, E_j] = -[E_j, E_i]$, which implies in particular $[E_i, E_i] = 0$.

It may be of interest to note that in the case of infinitesimal rotations $E_4, E_5, E_6, [E_i, E_j]$ is simply the outer or vector product $E_j \times E_i$. Again, if E_i and E_j (or equivalently, $\exp(tE_i)$ and $\exp(uE_j)$) are *permutable*, so that $E_i E_j = E_j E_i$, then $[E_i, E_j] = 0$, and conversely.

It will be noted that in (35), we have always

$$(36) \qquad [E_i, E_j] \doteq \sum_k c_k^{ij} E_k$$

for suitable constants c_k^{ij}. It is a fundamental theorem of Lie, that relations like (36) hold for *any* finite-parameter group. The constants c_k^{ij} are called the *structure constants* of the group, and determine the group to within isomorphism.

I hope that the explanations which I have just given will make the results derived below understandable, even though the derivations cannot be understood except by those already familiar with Lie theory.

---

[27] I use throughout the convention that $E_i E_j$ means "first E_i, then E_j."

112. Forces and commutators

Now let G be an arbitrary r-parameter Riemannian group manifold, and let C be any one-parameter subgroup of G, generated by the infinitesimal transformation E.

One can always ([63], p. 47) introduce *canonical parameters* into G near the identity, with a basis $E_1, E_2, \cdots, E_n$ of infinitesimal transformations, so that if $q = (q_1, \cdots, q_n)$ is any sufficiently small vector element of G,

$$(37) \qquad q = \exp(q_1 E_1 + \cdots + q_n E_n).$$

From this generalization of (32) one can derive the following generalization of (31),

$$(38) \qquad \lambda q \cdot \mu q = (\lambda + \mu) q,$$

where $r \cdot s$ denotes the group product of r and s in G.

Now consider the geodesic curvature of C at $q = 0$, relative to the metric $ds^2 = \sum T_{ij} dq_i dq_j$. By §108, this is proportional to

$$(39) \quad Q_i = \frac{d}{dt}\left(\frac{\partial T}{\partial \dot{q}_i}\right) - \frac{\partial T}{\partial q_i} = \frac{d}{dt}(T_{ik}\dot{q}_k) - \frac{1}{2}\frac{\partial T_{hk}}{\partial q_i}\dot{q}_h\dot{q}_k$$

$$= T_{ik}\ddot{q}_k + \dot{q}_k\frac{\partial T_{ik}}{\partial q_h}\dot{q}_h - \frac{1}{2}\frac{\partial T_{hk}}{\partial q_i}\dot{q}_h\dot{q}_k = \frac{\partial T_{i1}}{\partial q_1} - \frac{1}{2}\frac{\partial T_{1i}}{\partial q_i},$$

since $\dot{q}_1 = 1$, $\dot{q}_j = 0$ for $j = 2, \cdots, r$, and $q_k = 0$ for all k. This is a disguised Christoffel tensor (13′); by §110, the constant of proportionality is $v^2 = T_{11}$, which we can reduce to unity by appropriate choice of time scale. Ultimately, it is Q and not the geodesic curvature which really interests us, so that the constant of proportionality need not trouble us.

We now compute the partial derivatives occurring in the last formula of (39). To do this, we note that by definition, the infinitesimal vector dq' issuing from $\exp(tE_i)$ is equivalent under left-translation to the infinitesimal vector dq issuing from the identity 0, if and only if

$$(40) \qquad \exp(tE_i + dq') = \{\exp(tE_i)\} \cdot dq.$$

But the Schur-Campbell-Hausdorff series[28] expresses the right hand side of (40) in the form

$$(41) \qquad \{\exp(tE_i)\} \cdot dq = \exp(tE_i + dq + \tfrac{1}{2}t[E_i, dq] + \cdots),$$

where the omitted terms are quadratic in t. Since dq and dq' are equivalent as regards their infinitesimal part, we infer

$$(41') \qquad dq = dq' - \tfrac{1}{2}t[E_i, dq'] + \cdots.$$

[28] This classical formula is proved under extremely general conditions as Theorem 14, on p. 92 of my paper "Analytical groups", *Trans. Am. Math. Soc.*, *43* (1938), 61–101.

Now writing $d\boldsymbol{q} = dq_1 E_1 + \cdots + dq_n E_n$ and $d\boldsymbol{q}' = dq_1' E_1 + \cdots + dq_n' E_n$, we get by definition, as in (36),

$$[E_i, d\boldsymbol{q}'] = dq_j'[E_i, E_j] = dq_j' c_h{}^{ij} E_h.$$

Substituting back into the vector equation (41'), and equating corresponding components, we obtain the fundamental relation

$$(42) \qquad dq_h = dq_h' - \tfrac{1}{2} t c_h{}^{ij} dq_j' + \cdots,$$

neglecting quadratic and higher terms in t.

But by the definition of Riemannian group manifolds, ds^2 is invariant under left-translations. Hence by (40) and (42),

$$
\begin{aligned}
dq_h' T_{hk}(tE_i)dq_k' &= dq_h T_{hk}(\mathbf{0})dq_k \\
&= (dq_h' - \tfrac{1}{2} t c_h{}^{ij} dq_j') T_{hk}(\mathbf{0})(dq_k' - \tfrac{1}{2} t c_h{}^{il} dq_l') \\
&= dq_h' T_{hk}(\mathbf{0})dq_k' - \tfrac{1}{2} t \{ c_h{}^{ij} dq_j' T_{hk}(\mathbf{0})dq_k' + c_k{}^{il} dq_h' T_{hk}(\mathbf{0})dq_l' \},
\end{aligned}
$$

neglecting quadratic and higher terms in t. Equating coefficients, by interchanging the "dummy indices" of summation j, h and l, k in curly brackets, we get

$$T_{kh}(tE_i) = T_{hk}(\mathbf{0}) - \tfrac{1}{2} t \{ c_j{}^{ih} T_{jk}(\mathbf{0}) + c_l{}^{ik} T_{hl}(\mathbf{0}) \} + \cdots.$$

Now differentiating with respect to t, and writing both summation indices as j, we get the formulas

$$(43) \qquad \frac{\partial T_{hk}}{\partial q_i} = -\tfrac{1}{2} \{ c_j{}^{ih} T_{jk} + c_j{}^{ih} T_{hj} \} \quad \text{at } \mathbf{0}.$$

Now substituting (43) into (39), we get

$$4Q_i = -2c_j{}^{1i} T_{j1} - 2c_j{}^{11} T_{ij} + c_j{}^{i1} T_{j1} + c_j{}^{i1} T_{1j}.$$

Moreover, by the well-known antisymmetry $[E_i, E_1] = -[E_1, E_i]$ of structure constants, $-c_j{}^{1i} = c_j{}^{i1}$ and $c_j{}^{11} = 0$. Substituting back and dividing by four, we get our final formula[29]

$$(44) \qquad Q_i = c_j{}^{i1} T_{1j}.$$

Obviously, we get the corresponding formula

$$(44^*) \qquad Q_i = \sum_{j=1}^{n} c_j{}^{ih} T_{hj} = -\sum_{j=1}^{n} c_j{}^{hi} T_{hj}$$

in the case of steady motion along the E_h-axis. Hence *steady motion of a particle on any Riemannian group manifold G under E_h requires the external force* (44*). Alternatively, $c_j{}^{ih} T_{hj}/T_{hh}$ (summed over j but not h) expresses

[29] Formula (44) was obtained independently in 1945 by John Breakwell and myself. See Abstract 52-7-242, *Bull. Am. Math. Soc.*, 52 (1946), p. 617.

the geodesic curvature of the one-parameter subgroup $\exp(tE_h)$ on G. If we choose a normal orthogonal basis $E_1, \ldots, E_n$, relative to the metric ds^2 at 0, this curvature is simply $c_h{}^{th}$.

113. Applications

From the single general formula (44*), and the special commutation relations (35) for the Euclidean group, one can deduce the external force required to maintain steady translation or steady rotation in an ideal fluid. (The force exerted by the fluid on the body is of course the reverse of this force.) Thus, under steady translation E_1 along the x_1-axis, the force required is

$$(45) \qquad\qquad (0, 0, 0; 0, T_{13}, -T_{12}).$$

Under steady rotation E_4 with angular velocity of one radian per second about the x_1-axis, the force is similarly,

$$(46) \qquad\qquad (0, T_{43}, -T_{42}; 0, T_{46}, -T_{45}).$$

These are the classical formulas of Kirchhoff and Kelvin,[30] it will be noted that (45) reduces to 0 (steady motion under no constraints), if the kinetic energy tensor $\sum_{i,j=1}^{3} T_{ij}\dot{q}_i\dot{q}_j$ for *translation* has been diagonalized. In other words, steady translation without external forces is (theoretically) possible along the characteristic axes of the translation kinetic energy tensor, and no others.[31]

Although the preceding formulas are highly theoretical, and steady translation in the absence of external forces is not physically possible, formula (45) does provide the classical explanation for the *broadsiding tendency of a flat plate*. Using (45), it is not hard to show that the *stable* steady translation is along the characteristic axis *maximizing* the kinetic energy tensor for translation. This result is in qualitative accord with experiment.

We may note, as another application of our general formula (44*), the fact that the component of generalized force is zero in any direction corresponding to an infinitesimal transformation which is *permutable* with the steady motion in question. For if $[E_h, E_i]=0$, then Q_i is zero.

This provides a group-theoretic motivation for the d'Alembert Paradox: steady translation along an axis gives rise (theoretically) to no translation thrust at all, but only to a couple, since all translations are permutable. Since translations and rotations about the same axis are permutable,

[30] See [7], p. 169, formula (4), for (45); (46) is implicit in [7], §125. See also the references given there.
[31] The stability of such translations has been considered by H. D. Ursell, *Proc. Camb. Phil. Soc.*, 37 (1941), 150–167.

translation also gives rise to no torque about the axis of translation; the moment axis is perpendicular to the axis of translation.

The same consideration leads immediately to what may be called the Propeller Paradox. Screw motion about an axis can (in the classical theory) give rise to no axial thrust *or* torque about the axis of the motion! Hence for a propeller, or other object possessing n-fold rotational symmetry about that axis $(n > 1)$, *all* force components are (theoretically) zero.[32]

As shown in [14], Chapter V, §14, the preceding results have generalizations to the case of imaginary rigid bodies in an ideal fluid filling a *non-Euclidean* space. However, we will not repeat these generalizations here, because their physical significance is too remote. The main point is, that *the classical theory of the motion of a rigid body through an ideal fluid can be treated as a topic in the Lie theory of homogeneous spaces.*[33]

114. Stokes damping

In the case of small oscillations, the Lagrangian analysis of the preceding sections predicts an added inertia, which will lengthen the period of free oscillations, but it predicts no damping. The first theoretical analysis of the damping of free oscillations, by *viscosity*, was made in 1850 by Stokes. This analysis neglects convection, which is reasonable for sufficiently small oscillations, and linearizes the equations of motion. Because of this linearization, it predicts a "logarithmic decrement" (defined as the logarithm of the ratio of the amplitude of successive oscillations) which is amplitude-independent. We briefly sketch the method of calculation.

After allowing for buoyancy, the linearized equations of motion reduce simply to

$$(47) \qquad \frac{\partial u}{\partial t} = -\frac{\nabla p}{\rho} + \nu \nabla^2 u.$$

We can eliminate p by taking the curl of both sides. Denoting the vorticity $\nabla \times u$ by ζ, we get

$$(48) \qquad \frac{\partial \zeta}{\partial t} = \nu \nabla^2 \zeta.$$

In the case of *forced* sinusoidal oscillations of constant amplitude, with angular frequency ω, (48) is equivalent to

$$(48^*) \qquad \nabla^2 \zeta = \left(\frac{i\omega}{\nu}\right)\zeta.$$

---

[32] Since the exterior of a propeller is simply connected, we cannot remedy the situation by introducing "circulation" about the propeller blades—without considerable "fudging".

[33] C. Chevalley, *Theory of Lie groups*, Princeton, 1946, p. 29 ff.; [63], Ch. V.

Here we adopt the usual convention that the *physical* vorticity is the *real* part of a complex function of space-time, complex analytic in time.

In the case of plane and axially symmetric flows (i.e. of cylinders oscillating transversely and solids of revolution oscillating along their axis), ζ can be expressed in terms of Stokes's stream function V. (Thus, in plane flows, $\zeta = \nabla^2 V$.) This greatly simplifies the boundary conditions.

We pass over the details of the calculations, which are long ([13], vol. 3, pp. 22–54) or ([7], pp. 632–45). For a *sphere* of radius a, the final dynamic effect is a force ([7], p. 644, (26))

$$(49) \qquad -X = \rho \text{ vol } (\mathbf{\Sigma}) \left\{ \frac{1}{2}\ddot{q} + \frac{9S}{2\sqrt{2}}\dot{q} + \frac{9}{2}\left(\frac{S}{\sqrt{2}} + S^2\right)q \right\}, \qquad S = \frac{1}{a}\sqrt{\frac{\nu}{\omega}}.$$

That is, besides the in-phase inertial induced mass $\frac{1}{2}\rho$ vol $(\mathbf{\Sigma})$ of §98, there is a sinusoidal force whose magnitude is $9S$ as large which is 135° out-of-phase, and one $9S^2$ as great which is 180° out-of-phase with the oscillations.

The last force is simply the resistance to steady "creeping flow" past a sphere moving with constant speed (§30). The former can be interpreted as *boundary-layer damping*, an idea which we shall now define in general.

115. Boundary-layer damping

In 1850, Stokes ([13], vol. 3, p. 21) suggested that "the effect of the fluid may be calculated with a very close degree of approximation, by regarding each element of the surface of the solid as an element of an infinite plane oscillating with the same linear velocity." Though Stokes only suggested this for the torsional oscillations of a solid of revolution about its axis, the same approximation has been suggested for small translational oscillations.[34] Since it is implied by Prandtl's boundary-layer theory (§27) when convection is neglected, we shall call the force calculated the *boundary-layer force*.

The calculation is easy for sinusoidal translation oscillations. The force is 135° out-of-phase with the motion ([7], p. 620). Hence, if F is the maximum force, and $-\dot{E}$ is the average rate of energy dissipation per unit time, then clearly $-\dot{E} = F\dot{q}/2\sqrt{2}$, where $\dot{q}$ is the maximum oscillation speed of the solid $\mathbf{\Sigma}$. But the average rate of energy dissipation on a surface element with area dS is ([7], p. 620, (9))

$$(50) \qquad -d\dot{E} = \rho\sqrt{\frac{\nu\omega}{8}}\, v_t{}^2(\mathbf{x})\, dS,$$

[33] J. Boussinesq, *J. de Math.*, *4* (1878), 335–76. A second-order correction has been proposed by G. F. Carrier and R. C. di Prima, *J. Appl. Mech.*, *23* (1956), 601–5.

where $v_t(x)$ is the maximum local tangential velocity in Euler flow. Knowing $v_t(x)$, one can calculate $-\dot{E}$ and hence F, by integrating (50) over the body surface.

Thus, if Σ is a sphere of radius a, then $v_t(x) = (3\dot{q}/2) \sin \theta$. Integrating, we get $-\dot{E} = 3\pi\rho\sqrt{\nu\omega/2}\ \dot{q}^2 a^2$. Consequently, the boundary layer force on a sphere is

$$(51) \qquad F = 6\pi\rho q\sqrt{\nu\omega} = [\tfrac{9}{2}\rho \text{ vol } (\Sigma)]\left(\frac{\sqrt{\nu\omega}}{a}\right) = 9m'S.$$

A similar formula may be derived for a solid of general shape in sinusoidal oscillation.[35]

By taking the Fourier transform of (48), one can also calculate the resulting force for a small motion having a general time history.[36] However, the big question is as to the *applicability* of the formulas so calculated; this we now consider.

116. Large-amplitude oscillations

Though pendulum clocks are less important today than in 1800, many experiments have been performed to check Stokes's formula (49).[37] It is clear that the added mass factor k and the damping coefficient are functions of relative amplitude α as well as of the Stokes number S. Unfortunately, α is variable in free damping, and is not recorded in most experiments; for these and other reasons, the significance of many experiments is obscure.

The most definitive experiments at large amplitudes and low S are perhaps those of Keulegan and Carpenter,[38] on cylinders and plates held in an oscillating liquid. Their data show a curious dependence of added mass and damping on relative amplitude, ignored by Stokes's formulas.

Let $x = A \sin \beta t$ denote the time-dependent displacement, so that $T = 2\pi/\beta$ is the period, $\theta = \beta t$ the phase, and $u_m = A\beta$ the maximum velocity. The force $X(\theta)$ was measured as a function of phase; by symmetry, $X(\theta + \pi) = -X(\theta)$. Let the diameter of Σ be d, so that the relative amplitude is $\alpha = 2A/d$.

[35] See M. J. Lighthill, *Proc. Roy. Soc.*, A224 (1954), 1–23.

[36] Rayleigh, *Phil. Mag.*, 21 (1911), 697–710; the formulas were derived earlier by Boussinesq and Basset.

[37] O. E. Meyer, *Jour. f. Math.*, 73 (1871), 31–68; M. I. Northway and A. S. Mackenzie, *Phys. Rev.*, 13 (1901), 145–64; G.F. Mc Ewen, *ibid.*, 33 (1911), 492–511; L. Marty, *J. de Phys. et Radium* (Paris), 6 (1935), 373–82; J. Valensi and C. Clarion, *Bull. Soc. France Mec.*, No. 8; E. G. Richardson and R. I. Tait, *Ost. Ing.-Archiv*, 8 (1954), 200–7. Other refs. are given below and in §§103–4; see also §§31–2.

[38] G. H. Keulegan and L. H. Carpenter, *J. Res. Nat. Bu. Standards*, 60 (1958), 423–40.

Measured values of $X(\theta)$ were fitted by the semi-empirical formula

$$(52) \qquad\qquad -X(\theta) \simeq c\dot{x}|\dot{x}| + M\ddot{x}.$$

Behind this formula was the idea that $-X(\theta)$ should be the sum of a

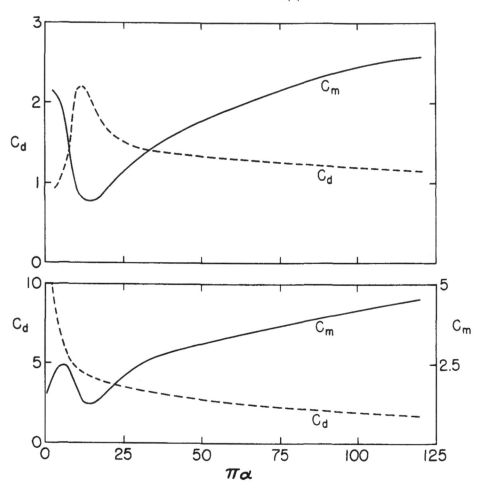

FIG. 28. Added mass of cylinder (above) and plate (below)

drag force $D = \frac{1}{2}\rho d\dot{x}^2 C_D$ proportional to the square of the velocity, and an inertial force $M\ddot{x} = (\pi\rho d^2/4)C_M\ddot{x}$ proportional to the acceleration. If this were exactly so, then C_D and C_M could be computed from the formulas

$$(53) \qquad\qquad C_D = \frac{-3}{4\rho u_m{}^2 d} \int_0^{2\pi} X(\theta)\cos\theta\, d\theta$$

and

(53')
$$C_M = \frac{4A}{\pi^2 \rho u_m^2 d^2} \int_0^{2\pi} X(\theta) \sin \theta \, d\theta.$$

Empirically, it was found that large-amplitude oscillation data were fitted quite well by (52), the empirical constants C_D and C_M being defined by (53)–(53'). Measured values of C_D and C_M depended primarily on the relative amplitude $\alpha = 2A/d$, and relatively little[39] on the Stokes number S. Graphs of the measured values of C_D and C_M are sketched in Fig. 28.

It is interesting to contrast formula (52) with that resulting from the small-amplitude approximations of Stokes, namely

(54) $-X(\theta) = c_1' \dot{x} + m^* \ddot{x}, \qquad m^* = (m' + m_0),$

m' being the ideal added mass and $m_0 = \rho \text{ vol }(\Sigma)$ the mass of fluid displaced. For a broadside flat plate or a circular cylinder, it is known ([7], p. 85) that $m' = \pi \rho d^2/4$. The term m_0 in (52') must be included because the obstacle is held at rest in an oscillating fluid; it would be absent if the obstacle were oscillating in a static fluid. One can compute the difference by noting that, in a liquid-solid mixture of constant density ρ, a sinusoidal translational oscillation of the whole exerts a force $m_0 \ddot{x}$ on the solid Σ. Obviously $X(\theta)$ is independent of ρ.

For comparison, the Stokes values $C_M = 1, 2$ (for cylinders and plates, respectively), deduced from the linearized theory, have also been plotted. It is obvious that Stokes's formulas are quite inapplicable to large oscillations.

117. Final conclusions

This book is intended as a contribution to natural philosophy. Borrowing from Newton,[40] I could say that "I offer this work as the mathematical principles of philosophy, for the whole burden of philosophy seems to consist in this—from the phenomena of motions to investigate the forces of nature, and from these forces to demonstrate the phenomena." I have tried to follow Aristotle's maxim:[41] "Let us first understand the facts, and then we may seek for the causes," being conscious also of Newton's maxim (*op. cit.*, p. 398): "We are certainly not to relinquish the evidence of experiments for the sake of dreams and vain fictions of our own devising."

[39] When $\alpha \gg 1$, correlations with the "Reynolds number" $\text{Re} = u_m d/\nu = \alpha/2S^2$ may be expected to be more important; see Bagliarello, *op. cit.* in §103.

[40] Preface to his *Principia Mathematica*, Motte translation as revised by F. Cajori, Berkeley, 1947, pp. xvii–xviii.

[41] Borrowed from E. T. Whittaker's brilliant monograph *Space and Spirit*, Edinburgh, 1946. Aristotle's views, which are often misstated, are given there on p. 27.

Accordingly, I began by describing many paradoxes to which purely dialectical arguments lead in the study of fluid motions, and the complexity of the real facts. This theme dominated the first three chapters, continued in the fourth, where some pitfalls of the conventional theory of modeling were exposed, and persisted intermittently to the very last sections. The novelty of this aspect of the presentation consists in its attribution of paradoxes to lack of *rigor*, a lack which is interpreted by some specialists as evidence of manly vigor—but which certainly would not have been condoned by Newton, Euler, Lagrange, Stokes, or any of the other founders of the science of fluid mechanics. A rigorous analysis of the theoretical foundations of real fluid mechanics constitutes, perhaps, the main critical contribution of this book.

I have chosen as my main positive contribution the applicability of the *group* concept to fluid mechanics. Thus, in the fourth chapter, this concept was revealed as the key to modeling theory ("similitude"). Many other applications of group theory were given in the fifth and sixth chapters. The culmination was an interpretation of the force on a rigid body in any steady motion through an ideal fluid as the curvature of the corresponding one-parameter subgroup in the Euclidean group space, metrized by the kinetic energy.

Thus, I have tried to throw two bridges across the widening gap between pure mathematics and physics. My success will have to be judged by the future work of those of my readers who find this book stimulating.

Bibliography

Chapters I and II

[1] G. Birkhoff, "Reversibility paradox and two-dimensional airfoil theory," *Am. Jour. Math.*, *68* (1946), 247–256.

[2] R. Courant and K. Friedrichs, *Supersonic flow and shock waves*, Interscience Press, 1948.

[3] C. Cranz, *Handbook of Ballistics* (translated), Vol. 1, His Majesty's Stationery Office, London, 1921.

[3a] H. Emmons, editor, *Fundamentals of Gas Dynamics*, Princeton University Press, 1958.

[4] S. Goldstein, editor, *Modern Developments in Fluid Dynamics*, 2 Vols., Oxford, 1938.

[4a] W. D. Hayes and R. F. Probstein, *Hypersonic Flow*, Academic Press, 1959.

[5] L. Howarth, editor, *Modern developments in fluid dynamics: high speed flow*, 2 Vols., Oxford, 1953.

[6] O. D. Kellogg, *Potential Theory*, Springer, 1929; reprinted by Dover.

[7] H. Lamb, *Hydrodynamics*, 6th ed., Cambridge, 1932. See also the Appendix of the second German edition, *Lehrbuch der Hydrodynamik*, Leipzig, 1931.

[8] L. M. Milne-Thomson, *Theoretical Hydrodynamics*, third edition, Oxford, 1955.

[9] C. W. Oseen, *Neuere Methoden und Ergebnisse in der Hydrodynamik*, Leipzig, 1927.

[10] K. Oswatitsch, *Gas dynamics*, English version by G. Kuerti, Academic Press, 1956.

[11] L. Prandtl and O. G. Tietjens, *Hydro- and Aeromechanics*, McGraw-Hill, 1934, 2 Vols.

[12] Rayleigh, J. W. Strutt Lord, *Scientific Papers*, 6 Vols., Cambridge University Press, 1899–1920.

[13] Stokes, Sir George, *Mathematical and physical papers*, Cambridge University Press, 1880–1905.

[14] G. Birkhoff, *Hydrodynamics: a study in logic, fact and similitude*, first ed., Princeton University Press, 1950.

Chapter III

[15] G. Birkhoff and E. H. Zarantonello, *Jets, Wakes and Cavities*, Academic Press, 1957.

[16] M. Brillouin, "Les surfaces de glissement de Helmholtz . . . ," *Ann. de Chim. Phys.*, *23* (1911), 145–230.

[17] P. R. Garabedian, H. Lewy and M. Schiffer, "Axially symmetric cavitational flow," *Annals of Math.*, *56* (1952), 560–602.

[18] D. Gilbarg, "Uniqueness of axially symmetric flows with free boundaries," *J. Rat. Mech. Anal.*, *1* (1952), 309–20.

[19] H. von Helmholtz, "Uber discontinuierliche Flussigkeitsbewegungen," *Monatsber. Berlin. Akad.* (1868), 215–28. Reprinted in *Phil. Mag.*, *36* (1868), 337–46 and in *Wiss. Abh.*, *1* (1882).

[20] C. C. Lin, *Theory of Hydrodynamical Stability*, Cambridge University Press, 1955.

[21] L. Prandtl and A. Betz, *Vier Abhandlungen zur Hydrodynamik und Aerodynamik*, Goettingen, 1927.

[22] Lord Rayleigh (J. W. Strutt), *Theory of Sound*, MacMillan, 2d ed., 1896, Chapters XX–XXI.

[23] H. Schlichting, *Boundary Layer Theory*, McGraw-Hill, 1955.

[24] R. V. Southwell and G. Vaisey, "Fluid motions characterized by free stream-lines," *Phil. Trans.*, *240* (1946), 117–61.

[25] G. K. Batchelor and R. M. Davies, eds., *Surveys in Mechanics*, Cambridge University Press, 1956.

[26] Sir Geoffrey Taylor, "The instability of liquid surfaces . . . ," *Proc. Roy. Soc.*, *201* (1950), 192–6.

[27] H. Villat, "Sur la résistance des fluides," *Ann. Sci. Ecole Norm. Sup.*, *28* (1911), 203–40.

[28] A. Weinstein, "Zur Theorie der Flüssigkeitsstrahlen," *Math. Zeits.*, *31* (1929), 424–33.

[29] G. Birkhoff, "Formation of vortex streets," *J. Appl. Phys.*, *24* (1953), 98–103.

[30] P. Garabedian, ". . . three dimensional cavities and jets," *Bull. Am. Math. Soc.*, *62* (1956), 219–35.

[30a] P. R. Garabedian, "Calculation of axially symmetric cavities and jets," *Pacific J. Math.*, *6* (1956), 611–84.

[31] D. Gilbarg and R. A. Anderson, "Influence of atmospheric pressure on . . . entry of spheres into water," *J. Appl. Phys.*, *19* (1948), 157–169.

[32] Kelvin, Sir W. Thomson, Baron, *Proc. Roy. Soc. Edinburgh*, 7 (1869), 384–390 and 668–672. See also *Math. and Phys. Papers*, Vol. 4, 69–75, 93–97, and 101–114.

[33] G. Kirchhoff, "Zur Theorie freier Flüssigkeitsstrahlen," *Crelle's Jour. für Math.*, *70* (1869), 289–298.

[34] J. Leray, "Les problèmes de représentation conforme de Helmholtz, théorie des sillages et des proues," *Commentarii Math. Helv.*, 8 (1935), 149–180 and 250–263.

[35] T. Levi-Civita, "Scie e leggi de resistenzia," *Rend. Circ. Mat. Palermo*, *23* (1907), 1–37.

[36] N. Levinson, "The asymptotic shape of the cavity . . . ," *Annals of Math.*, *47* (1946), 704–730.

[37] D. Riabouchinsky, *Proc. Lond. Math. Soc.*, *19* (1921), 206–215. See also *ibid.*, *25* (1926), 185–194, and E. G. C. Poole, *ibid.*, *25* (1926), 195–212.

[38] A. M. Worthington, *A Study of Splashes*, Longmans, 1908.

[39] F. S. Sherman, ed., "Naval Hydrodynamics," Vol. 1, *Publ. 515*, National Acad. Sci.—Nat'l Res. Council, Washington, 1957.

[40] Ralph D. Cooper, ed., "Naval Hydrodynamics," vol. 2, Publication ACR–38, U.S. Govt. Printing Office, 1960.

[41] *Cavitation in Hydrodynamics*, H.M. Stationery Office, London, 1936.

Chapter IV

[42] G. Birkhoff and S. MacLane, *Survey of Modern Algebra*, rev. ed., New York, 1953.

[43] P. W. Bridgman, *Dimensional Analysis*, 2d ed., New Haven, 1931.

[44] J. M. Burgers, E. Dobbinga, L. Troost, and M. J. Bossen, *Modelproeven en Kengrootheden op Stromingsgebied*, Amsterdam, 1947.

[45] E. Buckingham, "On physically similar systems," *Physical Review*, *4* (1914), 354–376.

[46] E. Buckingham, "Model experiments and the forms of empirical equations," *Trans. Am. Soc. Mech. Eng.*, *37* (1915), 263–296.

[47] A. Craya, "Similitude des modèles fluviaux à fond fixe," *La Houille Blanche* (Grenoble), *3* (1948), 335–346.

[48] Mrs. T. Ehrenfest-Afanassjewa, "Dimensionsbegriff und Bau physikalischer Gleichungen," *Math. Annalen*, 77 (1915), 259–276.

[49] Mrs. T. Ehrenfest-Afanassjewa, "Dimensional analysis . . . theory of similitudes," *Phil. Mag.*, *1* (1926), 257–272.

[50] J. B. J. Fourier, *Théorie analytique de la chaleur*, 1822, English translation by A. Freeman, Cambridge Univ. Press, 1878.

[51] H. L. Langhaar, *Dimensional analysis and theory of models*, Wiley, 1951.

[52] A. Martinot-Lagarde, *Analyse dimensionelle. Applications à la mécanique des fluides*, Lille, 1946 (ozalid copies); printed in 1948 by the Groupement des Recherches Aéronautiques.

[52a] A. Martinot-Lagarde, "Similitude physique : exemples d'applications a la mécanique des fluides," Memorial 66 des Sciences Physiques, Gauthier-Villars, 1960.

[53] A. W. Porter, *The Method of Dimensions*, Methuen, 1933.

[54] Osborne Reynolds, *Collected Papers*, esp. *Vol. I*, 257–390, and *Vol. II*, 51–105.

[55] A. E. Ruark, "Inspectional Analysis: A method which supplements dimensional analysis," *Jour. Elisha Mitchell Sci. Soc.*, *51* (1935), 127–133.

[56] L. I. Sedov, *Similarity and dimensional methods in mechanics*, 4th ed., Moscow, 1957. (English translation, Academic Press, 1960.)

[57] J. J. Stoker, *Water waves*, Interscience Publ., 1957.

Chapter V

[58] K. Bechert, "Differentialgleichungen der Wellenausbreitung in Gasen," *Annalen der Physik, 39* (1941), 357–372; "Ebene Wellen in idealen Gasen mit Reibung und Warmeleitung," *ibid., 40* (1941), 207–248.

[59] G. Birkhoff, "Dimensional analysis and partial differential equations," *Electrical Engineering, 67* (1948), 1185–1188.

[60] G. D. Birkhoff, "The principle of sufficient reason," Rice *Institute Pamphlet, 28* (1941), No. 1, pp. 24–50.

[61] H. Blasius, "Grenzschichten in Flussigkeiten mit kleiner Reibung," *Zeits. Math. Phys., 56* (1908), 1–37.

[62] L. E. Dickson, "Differential equations from the group standpoint," *Annals of Math., 25* (1924), 287–378.

[63] L. P. Eisenhart, *Continuous Groups of Transformations*, Princeton University Press, 1933.

[64] V. M. Falkner and S. W. Skan, "Solutions of the boundary layer equations," *Phil. Mag., 12* (1931), 865–96.

[65] Th. Geis, "Ahnlichen dreidimensionalen Grenzschichten," *J. Rat. Mech. Anal., 5* (1956), 643–86.

[66] L. Prandtl, "Neuere Untersuchungen über die strömende Bewegungen der Gase und Dämpfe," *Phys. Zeits., 8* (1907), 23–31.

[67] R. M. Redheffer, "Separation of Laplace's equation," Doctoral Thesis, Massachusetts Institute of Technology (1948). See also N. Levinson, B. Bogert, and R. Redheffer, "Separation of Laplace's Equation," *Quar. Appl. Math., 7* (1949), 241–262.

[68] A. Rosenblatt, "Solutions exactes des équations du mouvement des liquides visqueux," *Mem. Sci. Math., 72* (1935), 63 pp.

[69] L. I. Sedov, "On unsteady motions of a compressible fluid," *Doklady U.R.S.S., 47* (1945), 91–93, and *52* (1946), 17–20.

[70] K. P. Staniukovitch, *Unsteady Motion of a Continuous Medium*, Moscow, 1955 (English translation, Pergamon Press, 1959).

[71] K. P. Staniukovitch, "On automodel solutions of equations of hydrodynamics possessing central symmetry," *Doklady U.R.S.S., 48* (1945), 310–312; *60* (1948), 1141–1144; and *64* (1949), 29–32, 179–181, and 467–470.

[72] G. I. Taylor and J. W. Maccoll, "The air pressure on a cone moving at high speeds," *Proc. Roy. Soc., A139* (1933), 278–311; J. W. Maccoll, "The conical shock wave formed by a cone moving at high speed," *ibid., A159* (1937), 459–72.

[73] E. T. Whittaker, *Analytical Dynamics*, 4th ed., Cambridge, 1937.

[74] H. Weyl, "The differential equations of the simplest boundary-layer problems," *Annals of Math., 43* (1942), 381–407.

Chapter VI

[75] H. L. Dryden, F. D. Murnaghan, and H. Bateman, "Hydrodynamics," *Bull. 84, Nat. Res. Council,* Washington, 1932.

[76] Kelvin, Lord, "Hydrokinetic solutions and observations," *Phil. Mag., 42* (1871), 362–380.

[77] G. Kirchhoff, "Über die Bewegung eines Rotationskorpers in einer Flussigkeit," *Crelle's Jour., 71* (1869), 237–273.

[78] H. Lamb, "On the forces experienced by a solid moving through a liquid," *Quar. Jour., 19* (1883), 66–70.

[79] M. Schiffer and G. Szegö, "Virtual mass and polarization," *Trans. Am. Math. Soc., 67* (1949), 130–205.

[80] G. I. Taylor, "The energy of a body moving in an infinite fluid, with an application to airships," *Proc. Roy. Soc., A120* (1928), 13–22, and 260–283.

[81] Th. Theodorsen, *Impulse and Momentum in an Infinite Fluid,* Th. von Karman Anniversary Volume, Calif. Inst. Technology (1941), 49–58.

[82] W. Thomson and P. G. Tait, *Principles of Natural Philosophy,* 2 vols., esp. Sec. 313ff.

Index